W0042858

THE SEMANTICS OF RELATIONSHIPS

Information Science and Knowledge Management

Volume 3

Editor-in-Chief:

J. Mackenzie Owen, *University of Amsterdam, Amsterdam*

Editorial Board:

E. de Smet, *Universiteit Antwerpen, Wilrijk*
Y. Fujiwara, *Kanagawa University, Hiratsuka*
M. Hedstrom, *University of Michigan, Ann Arbor, MI*
A. Klugkist, *UB Groningen, Groningen*
K.-D. Lehmann, *Stiftung Preussischer Kulturbesitz, Berlin*
C. Lupovici, *Université de Marne la Vallee, Marne la Vallee*
A.M. Paci, *Istituto di Studi Sulla Ricerca e Documentazione Scientifica, Roma*
M. Papazoglou, *Katholieke Universiteit Brabant, Tilburg*
D.J. Waters, *The Andrew W. Mellon Foundation, New York*

The titles published in this series are listed at the end of this volume.

THE SEMANTICS
OF RELATIONSHIPS

An Interdisciplinary Perspective

edited by

REBECCA GREEN

College of Information Studies,
University of Maryland, College Park, Maryland, U.S.A.

CAROL A. BEAN

Extramural Programs,
National Library of Medicine, Bethesda, Maryland, U.S.A.

and

SUNG HYON MYAENG

Natural Language and Information Retrieval Laboratory,
Division of Information and Communication,
College of Engineering, Chungnam National University, Taejon, Korea

KLUWER ACADEMIC PUBLISHERS
DORDRECHT / BOSTON / LONDON

A C.I.P. Catalogue record for this book is available from the Library of Congress.

ISBN 978-1-4020-0568-8

Published by Kluwer Academic Publishers,
P.O. Box 17, 3300 AA Dordrecht, The Netherlands.

Sold and distributed in North, Central and South America
by Kluwer Academic Publishers,
101 Philip Drive, Norwell, MA 02061, U.S.A.

In all other countries, sold and distributed
by Kluwer Academic Publishers,
P.O. Box 322, 3300 AH Dordrecht, The Netherlands.

Printed on acid-free paper

All Rights Reserved
© 2002 Kluwer Academic Publishers
No part of this work may be reproduced, stored in a retrieval system, or transmitted
in any form or by any means, electronic, mechanical, photocopying, microfilming, recording
or otherwise, without written permission from the Publisher, with the exception
of any material supplied specifically for the purpose of being entered
and executed on a computer system, for exclusive use by the purchaser of the work.

Table of Contents

Introduction

The genesis of this volume was the participation of the editors in an ACM/SIGIR (Association for Computing Machinery/Special Interest Group on Information Retrieval) workshop entitled "Beyond Word Relations" (Hetzler, 1997). This workshop examined a number of relationship types with significance for information retrieval beyond the conventional topic-matching relationship. From this shared participation came the idea for an edited volume on relationships, with chapters to be solicited from researchers and practitioners throughout the world. Ultimately, one volume became two volumes. The first volume, *Relationships in the Organization of Knowledge* (Bean & Green, 2001), examines the role of relationships in knowledge organization theory and practice, with emphasis given to thesaural relationships and integration across systems, languages, cultures, and disciplines. This second volume examines relationships in a broader array of contexts. The two volumes should be seen as companions, each informing the other.

As with the companion volume, we are especially grateful to the authors who willingly accepted challenges of space and time to produce chapters that summarize extensive bodies of research. The value of the volume clearly resides in the quality of the individual chapters.

In naming this volume *The Semantics of Relationships: An Interdisciplinary Perspective*, we wanted to highlight the fact that relationships are not just empty connectives. Relationships constitute important conceptual units and make significant contributions to meaning. We also wanted to highlight the parallels that exist between work on relationships in various communities, many of which are interdisciplinary in their own right. The specific disciplines and specializations with which the individual authors identify themselves are noted in the List of Contributors, which follows this Introduction.

The chapters in the volume have been organized into three parts. The first part explores types of relationships; the second delves into the role of relationships in knowledge representation and reasoning; the third presents applications that make central use of relationships. None of these three sections pretends to be complete in its coverage (nor are they self-contained): The chapters included are only representative—indeed perhaps only suggestive—of all that deserves discussion.

TYPES OF RELATIONSHIPS

A ubiquitous and timeless human cognitive activity is the ongoing effort each individual makes to construct a cohesive and predictable mental view of the world around her, seeking patterns by which to organize and make sense of it. This involves conceptually clustering things and ideas into named categories based on observable shared characteristics judged salient in a given context, with the resulting categories held together by some sort of mental framework of relationships; in short, classification. Our understanding of the world then, as well as our ability to survive in it, depends crucially on our innate ability to perceive and characterize the relationships between concepts, that is, to construct conceptually valid and robust classes of concepts and the relationships among them. Conceptual structuring relationships are thus an integral part of the very foundation on which we build and organize our knowledge and understanding of the world in which we live. If concepts are seen as the basic building blocks of conceptual structure, then relationships are the mortar that holds it together. However, relationships have received but a small portion of attention compared to what has been devoted to the concepts and concept classes themselves.

In addressing specific types of relationships, the first section focuses on the most basic conceptual structuring relationships, or more accurately as will be seen, relationship types. Three primary meta-classes of relationships are widely recognized: relationships of equivalence, of hierarchy, and of association. Relationships in each category contribute to structuring our conceptual framework. Equivalence relationships appear in this volume as the identity relationships of chapter 7 (along with hierarchical subsumption relationships), but are addressed more generally as a recurring theme throughout the companion volume described above.

There seems to be little argument that our most common conceptual structures are basically hierarchical in nature. These are the best known, most studied, and therefore best understood of all structuring relationships. Three types of hierarchical relationships are discussed in detail in the first three chapters in this section. Thus our exploration of relationship types in this volume begins with the prototypical structuring relationship, that of hyponymy or inclusion, often referred to, variously, as the is-a (isa, IS-A, ISA) relationship. Next we examine troponymy, a class of relationships similar to hyponymy, but which instead deals with verbs and processes rather than nouns (or noun phrases) and entities. With meronymy, we return to the other fundamental ontological relationship obtaining among nouns and entities, that of wholes and parts. Hyponymy has received the lion's share of attention, but meronymy may be an even more fundamental means of making sense of the world around us. As the chapters in this section demonstrate, the richness of even these most elemental relationships provides some hint of the complexity facing us in our attempt to harness and exploit the power of this semantic device.

Although there have been a number of efforts to compile inventories of associative relationships, there is surprisingly little agreement on the details or widespread acceptance of a canonical subset of this third relationship type. Because an in-depth treatment of the vast range of associative relationship types is far beyond the scope and scale of this volume, the fourth chapter in part I provides a close look at just a single type of associative relationship, but one that is particularly important in human cognition, the cause-effect relationship. The primacy of this relationship is reflected in its role in making sense of our past interactions with the world and in our ability to predict the course and outcome of future interactions.

The paradigmatic sense relation of hyponymy among nouns is addressed by Alan Cruse in chapter 1. This seemingly straightforward relationship of inclusion is nearly ubiquitous in human conceptual structures of all kinds, but efforts to characterize it precisely have been unsuccessful for several reasons. Arguing that idealized definitions of hyponymy are over-strict and fail to capture the intuitions of native speakers, which suggests the existence of 'natural categories' of hyponyms, Cruse explores the characterization of hyponymous pairs using prototype theory, but this approach also falls short of satisfying. An important variety of hyponymy is taxonomy (is-a-kind-of), as distinct from 'simple' hyponymy (is-a):

This is the vertical structuring relationship of taxonomic hierarchies. A taxonym and its hypernym clearly must share the same 'perspective,' but it remains unclear just how that shared 'essence' is specified and whether the notion of 'essence' is completely distinct from the 'highlighting' of salient features. Other dimensions of hyponymic variation correlate with the taxonomic level of related items and whether the specification of a hyponym *vis-à-vis* its hypernym relies upon single-feature or encyclopedic characterization.

Hyponymous relationships occur in all the major and in many of the minor syntactic categories. Christiane Fellbaum examines the semantics of troponymy, a class of hyponymy-like hierarchical relationships among verbs, in chapter 2. Instead of specification by kind, troponymy relates verb pairs by elaborating manner for the subordinate member relative to the superordinate member. In this fashion, troponymy appears to be the primary structural device for conceptual organization of verb meanings. Fellbaum discusses the consequences of encoding a manner component in the verb lexicon and draws distinctions among different kinds of troponymy. Surprisingly, semantic verb classes vary in the syntactic behavior or characteristics exhibited at different levels of manner elaboration, in some cases according syntactic privileges to the superordinate (more elaborated) verb and in others to the subordinate (unelaborated) verb in a pair.

The decomposition of cognitively complex concepts can result in two distinct structural forms, depending on the method of partitionment chosen. Taxonomies result from creating a framework of subclasses, while separating the larger whole into distinctive parts or portions yields partonomies, or meronomies (Tversky, 1990). In chapter 3, Simone Pribbenow addresses the class of relationships associated with the latter: meronymy. While there are similarities in the knowledge structures resulting from these two types of relationships, there are clear distinctions between them that have important implications for knowledge organization and representation, particularly inclusion and inferencing processes. After examining cognitive aspects of parts, that is, their role in human conceptual and perceptual systems, Pribbenow discusses the fundamental approach to formalization of parts, classical extensional mereology. The shortcomings revealed when its axioms are applied to common-sense understanding of part-whole relations lead to a constructive approach to classification that can also accommodate more complexity in such relationships.

"Knowledge of cause and effect provides the basis for rational decision-making and problem-solving" (p. 51), thus governing a tremendous amount of human thought and activity. As such, human understanding and use of relationships of causation has received the attention of philosophers and psychologists for millenia. A broad overview of the cause-effect relationship is presented in chapter 4 by Christopher Khoo, Syin Chan, and Yun Niu, with particular attention to its textual expression. Causation can be explicitly and implicitly expressed using such linguistic constructions as causal links, causative verbs, resultative constructions, conditionals, and causative adverbs, adjectives, and prepositions. In addition, certain experiential and action verbs possess "causal valance," inherently biasing the attribution of causality to participants occupying a particular syntactic role. Modeling the complexities of causal knowledge structures will require developing typologies of cause as well as distinguishing the various aspects or roles for different types of causation or causal situations.

RELATIONSHIPS IN KNOWLEDGE REPRESENTATION
AND REASONING

A major reason why we care about relationships is the criticial role they play in how we represent knowledge—psychologically, linguistically, and computationally. Many systems of knowledge representation start with a basic (and explicit) distinction between entities and relationships, reflected, for example, in:

- the nodes and arcs of semantic networks,
- Langacker's distinction between
  - nominal predications (which designate things and correspond to nouns) and
  - relational predications (which designate states—"atemporal relations"—and processes, and correspond to adjectives, adverbs, prepositions, and verbs) (Langacker, 1987, p. 183), and
- the terms and propositions of first-order predicate calculus.

It is the presence of relationships at the most basic level of knowledge representation that allows higher-level (i.e., more complex) entities to be generated. The nature of the higher-level entities that emerge depend, at least in part, on the semantic content of the relationships, as explored in part I of the book.

But the importance of relationships is not limited simply to how we represent knowledge, but also to how we reason with it. If we know that A and B are related as class and instantiation we will reason about them in a completely different way than if we know they are related as cause and effect. Such relational knowledge is a key element in how we reason about our world.

The increased understanding of the richness of relationships addressed in part I leads us to examine in part II the roles relationships play in human perception of the world and our interaction with it, on the one hand, and the roles that relationships can play in more formal systems that represent and reason with knowledge, on the other.

In chapter 5, Rebecca Green visits the role of relational structure in human thought, perception, reasoning, and imagination—"human knowledge representation and reasoning"—from the perspective of cognitive semantics. Scholars in this arena view such phenomena as image schemata, basic-level concepts, and frame semantic structures, all of which have internal structure, as basic conceptual units. For the cognitive semanticist, structured models organize our conceptual world at seemingly every turn. This view contrasts with a common representational assumption that our knowledge of the world is composed ultimately of conceptually simple entities and relationships between or among them. This latter view does not deny that almost any entity *could* be seen as being composed of more primitive parts, but it assumes that whatever we conceive as basic entities are regarded *as if* they had no internal structure. The cognitive semantics viewpoint counters that internal structure is a cardinal characteristic of basic entities, not simply in reality, but also in our perception. For example, our most basic concepts—among which we include such image schemata as containers, paths, forces, scales, and collections—generally involve a small number of parts in well-established configurations. The relational nature of the configuration is the basis of the concept. Likewise, our choice between several hierarchically-related terms (e.g., *furniture, chair, rocking chair*) often coincides with the hierarchical level in which a well-established part-whole configuration first emerges. The evidence suggests that entities don't just enter into relationships; they comprise relationships.

Second-order conceptual structures, for example, metonymy, conceptual integration (blended spaces), and metaphor and analogy, involve mappings between components of more basic, underlying structures. For example, in the case of metonymy, the mapping creates links between entities that are related to each other in the real world, as when we speak of a set of literary works by the name of their author (*Have you read any good Shakespeare recently?*); in the case of metaphor, the mapping projects structure from one domain into another, actually creating the perception of structure in an abstract domain based on the perception of structure in a concrete domain. It is the internal structuring of basic conceptual phenomena that make such second-order conceptual structures possible. Simply put, the primary components of thought and also language cannot be comprehended without recourse to structure and relationships.

The second chapter in this section (chapter 6), by Eduard Hovy, is set in the context of systems of semantic relations, organized into ontologies. The first half of the chapter proposes a structured set of features that can be used to compare ontologies, based on their general characteristics. The intent of the proposal is to serve the need for a standardized way of representing the basic traits and qualities of an ontology, not only to compare ontologies with each other, but also to aid in evaluating them. Hovy proposes a rich and hierarchically structured set of features, organized under the three general aspects of form, content, and usage. The hierarchical organization consistently gets into a depth of three and four levels of features; all together, Hovy proposes nearly fifty characteristics useful in describing ontologies. This structured feature set essentially represents a metadata proposal for ontologies; like other metadata standards, it would ultimately need to address questions of vocabulary control, so that features could be reported in consistent ways across a set of descriptions. Without this, comparison would be made more difficult.

The second half of Hovy's chapter addresses a methodology by which ontologies can be compared with each other in greater detail, focusing on particular differences of their content and structure. Such detailed comparisons could aid in knowledge transfer to new domains and in establishing better interoperability between systems. A first problem in trying to compare how two ontologies treat any concept they may have in common is the need to identify which concepts in the two systems are (near-) equivalents of each other. Given the number of concepts that may be involved, an automated process for identifying pairs of (near-)equivalent concepts is clearly called for. Hovy proposes an algorithm for cross-ontology alignment, which has been tested on aligning (separately) the topmost regions of CYC and MIKROKOSMOS with top-level concepts in SENSUS. The iterative alignment/integration technique includes (1) heuristics for matching on name, definition, and taxonomic structure, which lead to initial alignment suggestions; (2) a function for combining alignment suggestions; (3) criteria for validating alignments; and (4) an evaluation metric.

While there are certainly practical uses for ontology alignment, the theoretical issues raised by some of the near misses are of no less import. Many words are polysemous, in that they have multiple, related senses. Unfortunately, there is little agreement on exactly what constitutes a sense; accordingly, ontologies (dictionaries, thesauri, etc.) may recognize fine-grained senses or rather broader-grained senses. Can two concepts be said to be equivalent if their granularity differs? What if the concepts are nearly identical but are seen to be subsumed by different concepts, as when *geisha* is taken to be a subset of adult females or is taken to be a kind of entertainer? What if the concepts highlight different aspects of a single phenomenon, as when *archipelago* is taken to be a group of islands or taken to be the expanse of water surrounding the group of islands? What if the concepts focus on closely related senses, as when *library* is taken to be a collection of documents or taken to be the building housing the collection? Such relationships are more or less subtle, but will require explication and systematic treatment if we are going to work integratively across multiple ontologies effectively.

In chapter 7, Nicola Guarino and Christopher Welty examine the ontological nature of the arguments of an *is-a* link, establishing the conditions under which such subsumption links are well-founded. Three notions play crucial roles in their discussion: identity, unity, and essence. Identity is perhaps the purest of equivalence relationships and pertains to conditions under which two descriptions are of the same thing. The unity notion grows out of the meronymic relationship and concerns the wholeness of objects that are constituted of parts. The concept of essence is based on whether a thing's properties can change over time and circumstance without affecting the thing's identity: A property is considered essential if it "necessarily holds" at any time and in every possible world. Guarino and Welty carefully establish a set of identity and unity conditions (ICs, UCs)—criteria by which we may establish if two things are the same and if a thing is a whole—some of which depend, in turn, on essential properties.

Toward the end of the chapter the discussion turns to consider the interaction of identity and unity with subsumption. Several constraints affecting subsumption are set forth, including a combined identity and unity disjointness constraint: "Properties with incompatible ICs/UCs are disjoint." Guarino and Welty then informally demonstrate, based on ten specific examples from CYC, Mikrokosmos, Pangloss, and WordNet, that some *is-a* relationships that initially seem plausible (e.g., "An organization is a group," "A person is both a physical object and a living being") may be incompatible with ontological commitments underlying identity and unity.

The ubiquity of polysemy is one indicator of the human mind's willingness to group (closely) related concepts together under an umbrella concept; distinctions between and among the member concepts are sometimes considered, sometimes ignored with some fluidity. Something akin to this phenomenon is at play in at least some of the examples given by Guarino and Welty. This leaves us with a sticky (but oft-recurring) question: Is it more important that our knowledge-based systems be clean and internally consistent (and thus more reusable) or that they reflect human conceptual processing? Both answers are attractive, but it is far from clear if the two answers are (or can be made to be) compatible. Guarino and Welty supply a set of tools that will be especially helpful if the simple, clean, reusable, and internally consistent option is chosen. But even if the option to mirror human reasoning is chosen, their criteria would still help focus attention on important aspects of conceptual modeling.

Chapter 8, the final contribution in this section, by Christophe Jouis, addresses the logic of relationships, which guides us in the direction of how we can reason with relationships. Jouis introduces a structured set of relationships situated within the cognitive level of Desclés' Applicative and Cognitive Grammar (ACG). This cognitive level comprises four types of primitives: elementary semantic types of entities (e.g., Boolean, individualizable, mass, distributive, collective, or place); operators used in forming more complex types (e.g., functions, defined by the semantic types of both their arguments and the results they return); fundamental static relationships; and fundamental dynamic relationships.

A scheme for some twenty basic static relationships is presented, organized into relationships of identification, relationships of differentiation, and relationships of disjointedness. In addition to delineating the functional specification of the argument and result types of a relationship, the scheme also defines the algebraic properties of specific types of relationships; for example, are they reflexive?; are they symmetric?; are they transitive? This rich backdrop enables the validation of instantiated conceptual representations by checking them for consistency: Are arguments and results of the correct semantic types? Do the defined algebraic properties of the relationships hold? Inasmuch as dynamic relationships involve passing from one static situation to another, much of the machinery set out for handling static relationships can also be pressed into service for dynamic relationships. Fortunately, since chains of conceptual representations would be cumbersome to validate manually, software for automated validation has been developed.

APPLICATIONS OF RELATIONSHIPS

It isn't difficult to find applications where relationships play a certain role, since both our mental and external worlds comprise different parts that are related to each other. As examples in the external world we find ecological systems, social systems, organizations, computer programs, etc. Our mental world—that is, knowledge—is in turn full of representations that correspond in salient ways to the external world. As relationships are essential to both knowledge of and representations of the external world, various tools and automated systems have been developed to utilize and manipulate them, which enjoy differing degrees of success. Symbol-based artificial intelligence systems are perhaps the best-known examples for making explicit use of relationships. Even the object-oriented paradigm from software engineering and databases cannot stand alone without relationships (although it contrasts in some ways with the relationship-oriented approach). Further, the explosion of the World Wide Web can be attributed to the idea of relating massively distributed digital documents.

While relationships are ubiquitous, the focus in this section is on applications where the interplay between information/knowledge and humans matters. Taking the particular viewpoint that knowledge means "the state and/or ability of linking and associating pieces of information in a meaningful way" (Myaeng et al., 2000), the applications addressed in the chapters of part III of this volume aim at building a system for knowledge processing or enhancing state-of-the-art information systems to the extent they become knowledge systems. More specifically, the four chapters individually describe one or more of the following aspects:

- explication of implicit relationships (identification and extraction),
- expression and representation of relationships,
- exploitation or direct use of relationships for the underlying applications, and
- exploration and discovery of relationships.

The goal of the research reported in these chapters is, of course, to enhance the performance of the underlying information systems and to aid human knowledge processing. More specifically, the first three chapters discuss how relationships can be explicated, expressed, and exploited for various application tasks. The last chapter addresses how visualization techniques can help increase the value of relationships while showing how relationships can be expressed in visual ways.

Figure 1 gives a schematic view of the interplay between relationships and functionality in the context of knowledge processing and applications that involve humans. Adding new relationships in knowledge representation helps enhance new functionality in the underlying tasks. From the reverse direction, new functionality promotes the use of relationships in knowledge representation of the underlying system. The relationships can be manipulated (i.e., identified, extracted, exploited, etc.) by means of a visual and/or linguistic interface to a human user or a system.

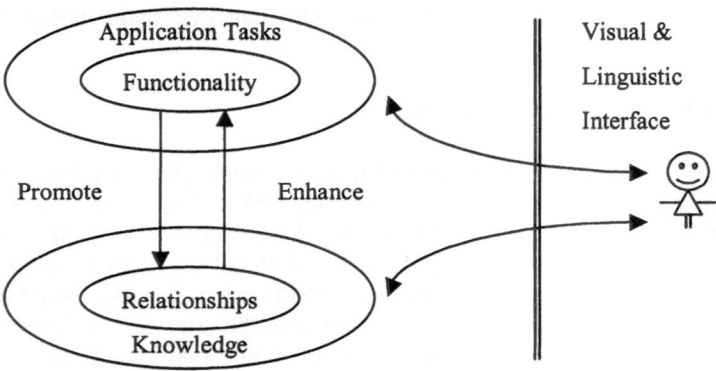

Figure 1. Interplay between relationships and application functionality

In chapter 9, Martha Evens deals with issues on the use of thesaural relations for information retrieval (IR) applications. She first surveys the types of paradigmatic relationships found in such resources as Roget's International Thesaurus and then describes two types of experiments in information retrieval research. One concerns the use of thesaural relations for the purpose of enriching queries and hence improving retrieval effectiveness. She ranges from early experiments with manual query expansion to those that expand queries automatically with such machine-readable resources as WordNet. This part focuses on how relationships can be *exploited* for IR. The other avenue Evens explores is the issue of constructing a thesaurus in an automatic way. The research efforts in this area began as an upshot of trends in information retrieval: elimination of human involvement in the IR process and the increasing scale of both IR experiments and IR practice. She introduces a variety of research efforts targeted at the use of machine-readable dictionaries and documents as the basis for extracting lexical-semantic relations. This part has to do with *explication* and linguistic *exploration* of relationships that exist in machine-readable resources.

In chapter 10, Chris Khoo and Sung Myaeng describe methods for *explicating*, *expressing*, and *exploring* syntagmatic relations, which are *exploited* for retrieval and extraction of information. Khoo and Myaeng first delve into a method for identifying relations using lexical patterns that can be expressed either linearly or graphically as conceptual graphs. Since linguistic patterns need to be identified as exhaustively and specifically as possible, they describe two different approaches to constructing such patterns from text. Specific methods and tools for automatic or semi-automatic means of pattern constructions are also introduced.

The second part of this chapter deals with how relations can be exploited in information retrieval and extraction applications. Unlike chapter 9, where thesaural or paradigmatic relations are of interest, this chapter focuses on syntagmatic relations that need to be identified from the text in documents. In the case of information retrieval, relational matching or structural matching is required to utilize the relationship information between two concepts. For information extraction applications, the main task of filling slots with textual values extracted from text is the process of identifying the relationship between the slot names and the textual values.

In chapter 11, Alexa McCray and Olivier Bodenreider describe the UMLS knowledge resources developed over many years at the National Library of Medicine for applications in the biomedical domain. With the ambitious goal of structuring and organizing a large number of biomedical concepts defined in various vocabularies into a coherent whole, the UMLS project has developed three knowledge sources: the Metathesaurus, the Semantic Network, and the SPECIALIST lexicon.

The Metathesaurus integrates more than fifty vocabularies or thesauri such as MeSH (Medical Subject Headings) and WHO's ICD (International Classification of Diseases). Concepts are explicitly related to multiple other concepts in various vocabularies with "ancestors" or "descendants" relationships. Furthermore, concepts can be related to each other by means of "Narrower" or "Broader" relationships or more refined relationships such as "location_of" that are used in the other knowledge source, the Semantic Network, which provides the general structure to which all Metathesaurus concepts are mapped. The SPECIALIST lexicon and related programs aid human specialists in dealing with various linguistic variations of terms and enrich the knowledge sources with natural language processing capabilities. This application includes all the four aspects of relationships mentioned above, supporting *explication* and *expression* for developers and facilitating *exploitation* and *exploration* for users and applications.

In chapter 12, Beth Hetzler describes how visualization and visual analysis can help users discover and explore relationships of various types. This chapter is unique among the chapters of this section, in that its emphasis is on how the underlying application (i.e., the visualization of information) can help increase the potential value of relationships. In other words, this is an instance in figure 1 of moving from application functionality to relationships.

Hetzler begins with a very informative introduction to the issues on visualization as well as motivations for visualizing numeric or textual information. With examples in metaphors and visual interaction techniques, the chapter describes the *expression* and *exploration* roles visualization techniques play. The rest of the chapter is devoted to seven different relationship types, for each of which at least an exemplary visualization technique is provided to demonstrate how the particular relationship type can be explored and discovered by human users.

As alluded to earlier, there are many other application areas where relationships play a crucial role. For instance, any application that belongs to the realm of knowledge-based systems would require explicit use of relationships, either typed or not. The hypertext field is another example where relationships serve as essential components for human-computer interactions as well as for existing applications like information retrieval (Joo & Myaeng, 1998; Kleinberg, 1998). Even in engineering design, relationships are used explicitly (Rucker & Aldowaisan, 1992). The applications presented here thus are representative of a much larger array of applications in which relationships play a prominent role.

References

Bean, C. A., & Green, R. (Eds.). (2001). *Relationships in the Organization of Knowledge* Dordrecht: Kluwer Academic Publishers.

Hetzler, B. (1997). Beyond word relations. *SIGIR Forum*, 31/2, 28-33.

Joo, W. K., & Myaeng, S. H. (1998). Improving retrieval effectiveness with hyperlink information. *Proceedings of the International Workshop on Information Retrieval with Asian Languages, IRAL '98*, 34-38.

Kleinberg, J. M. (1998). Authoritative sources in a hyperlinked environment. *Proceedings of the Ninth Annual ACM-SIAM Symposium on Discrete Algorithms*, 668-677.

Langacker, R. W. (1987). *Foundations of Cognitive Grammar (*Volume I: *Theoretical Prerequisites)*. Stanford: Stanford University Press.

Myaeng, S. H., Lee, M. H., Kang, J. H., Cho, E. I., Lee, Y. B., Lim, D. S., Lim, J. M., Oh, H. J., & Yang, J. S. (2000). A digital library system for easy creation/manipulation of new documents from existing resources. *Proceedings of the RIAO Conference*, 196-207.

Rucker, R,. & Adlowaisan, T. A. (1992). A design approach for constructing engineering scenario maps. *Computers & Mathematics with Applications* 23 (6-9), 419-440.

Tversky, B. (1990). Where partonomies and taxonomies meet. In S. L. Tsohatzidis (Ed.), *Meanings and Prototypes: Studies in Linguistic Categorization*, 334-344. New York: Routledge.

Rebecca Green
College of Information Studies
University of Maryland
College Park, Maryland, USA

Carol A. Bean
Extramural Programs
National Library of Medicine
Bethesda, Maryland, USA

Sung Hyon Myaeng
Natural Language & Information Retrieval Laboratory
Division of Information and Communication
College of Engineering
Chungnam National University
Taejon, Korea

List of Contributors

Olivier Bodenreider (medicine, knowledge representation)
Lister Hill National Center for Biomedical Communications
National Library of Medicine

Syin Chan (image/video compression, multimedia information retrieval)
Centre for Advanced Information Systems
School of Computer Engineering
Nanyang Technological University

D. Alan Cruse (lexical semantics, cognitive linguistics, pragmatics.)
Department of Linguistics
University of Manchester

Martha Evens (computer science, natural language processing)
Computer Science Department
Illinois Institute of Technology

Christiane Fellbaum (linguistics)
Cognitive Science Laboratory
Department of Psychology
Princeton University

Rebecca Green (knowledge organization, cognitive linguistics)
College of Information Studies
University of Maryland

Nicola Guarino (computer science, knowledge representation, formal ontology)
Institute for Systems Science and Biomedical Engineering of the Italian National
    Research Council (LADSEB-CNR)

Beth Hetzler (human computer interaction, information visualization, text mining)
Pacific Northwest National Laboratory

Eduard Hovy (computational linguistics, artificial intelligence, computer science)
Information Sciences Institute
University of Southern California

Christophe Jouis (knowledge organization, information retrieval, language
  resources, knowledge extraction and acquisition from corpora)
Université Paris - Sorbonne Nouvelle &
CAMS (Centre d'Analyse et de Mathématiques Sociales)
CNRS, EHESS, Université Paris - Sorbonne

Christopher Khoo (library & information science, information retrieval, information
  extraction, natural language processing)
Centre for Advanced Information Systems
School of Computer Engineering
Nanyang Technological University

Alexa T. McCray (linguistics, knowledge representation)
Lister Hill National Center for Biomedical Communications
National Library of Medicine

Sung Hyon Myaeng (computational linguistics, information retrieval)
Natural Language & Information Retrieval Laboratory
College of Engineering
Chungnam National University

Yun Niu (computational linguistics)
Centre for Advanced Information Systems
School of Computer Engineering
Nanyang Technological University

Simone Pribbenow (computer science, artificial intelligence, cognitive science)
Technologie-Zentrum Informatik (TZI)
Universität Bremen

Christopher Welty (computer science, knowledge representation, formal ontology,
  descriptive logics)
Computer Science Department
Vassar College

PART I

Types of Relationships

Chapter 1

Hyponymy and Its Varieties

D. Alan Cruse
Department of Linguistics, University of Manchester, Manchester, UK

Abstract:
This chapter deals with the paradigmatic sense relation of hyponymy as manifested in nouns. A number of approaches to the definition of the relation are discussed, with particular attention being given to the problems of framing a prototype-theoretical characterization. An account is offered of a number of sub-varieties of hyponymy.

1. INTRODUCTION

Hyponymy and its natural partner, incompatibility, are described by Lyons as "the most fundamental paradigmatic relations of sense in terms of which the vocabulary is structured" (1968, p. 453). Of all sense relations, they certainly occur across the widest range of grammatical categories and content domains. Hyponym-hyperonym pairs can be observed in all the major syntactic categories (more rarely in the minor categories). But a short time with a dictionary will be sufficient to convince anyone that it is much easier to come up with noun hyponyms than other types. The fact that well-developed taxonomic lexical hierarchies are almost exclusively nominal also reinforces the suspicion that the noun category is the natural home of hyponymy. Since everything that is interesting about hyponymy can be illustrated from the noun domain, the present chapter will deal exclusively with noun hyponyms.

Although this chapter focuses on hyponymy, it should be borne in mind that in the major function of articulating semantic space, hyponymy works in tandem with incompatibility. Hyponyms do not occur in isolation: A hyponym always has at least one incompatible co-hyponym; furthermore, it is possible that some varieties of hyponymy may be characterized in terms of relations between co-hyponyms (see especially the discussion of taxonomy below).

A number of different approaches to hyponymy will be considered, but ultimately the views expressed here are those of a lexical semanticist with a 'cognitive linguistic' bias.

2. WHAT DO SENSE RELATIONS RELATE?

An important basic question that must be faced at the outset is this: Hyponymy is a relation of meaning, but what are the entities that it relates? There are actually two questions here. The first concerns the 'size' of the meaning units involved; the second concerns their ontological status.

2.1 Upper and Lower Size Limits

I presume it may be taken as uncontroversial that hyponymy involves some sort of 'inclusion'. This can, of course, be interpreted in various ways: We can say, for instance, that the class of apples is included in the class of fruit, or we can say that the meaning of *fruit* is included in the meaning of *apple*. This naive notion of inclusion can be applied to units of any size. So, in an obvious way, the meaning of *John entered the room and sat down* 'includes' the meaning of *John sat down*; in a slightly less obvious way, the meaning of *John killed the wasp* includes the meaning of *The wasp died* (alternatively, instances of wasp-killing-by-John are a subclass of instances of wasp-dying.) There would therefore seem to be grounds for saying that *John killed the wasp* is hyponymous to *The wasp died*. However, this is not standard practice: The above type of sentential relation would normally be called entailment; we shall follow the standard practice and regard the largest units that can be related by hyponymy as being phrases. Notice that although the phrase *in the house* includes the meaning of *the house*, we would not say that the former is hyponymous to the latter. This is because hyponymy is a paradigmatic relation, which means that the related items must fit into the same grammatical slot, or, probably more fundamentally, they must be of the same semantic type (e.g., in categorial grammar terms).

When we look for a possible lower size limit for hyponymously related items, the picture becomes less clear. In one sense, there is no problem: We cannot go below morphemes, or more precisely, minimal semantic constituents, that is to say, the smallest units that are combined compositionally. However, a minimal semantic constituent may have a variety of readings in different contexts, and we need to ask whether these are to be treated jointly or individually in respect of sense relations, and if the latter, what the principle of individuation is to be.

In spite of Lyons's advocacy of lexemes as the proper relata of sense relations, it is generally accepted that the distinct senses of a polysemous lexeme are better treated separately. However, it is arguable that in some cases even senses should be subdivided for relational purposes. This is not the place to argue the matter in detail, but two types of case might be signaled. The first concerns what in Cruse (1995) are called facets, prototypical examples of which are the "physical object" and "abstract text" readings of *book* (called in Cruse, 1995 the [TOME] and [TEXT] facets, respectively). These have a high degree of autonomy and distinctness, one aspect of which is arguably their independent sense relations. At first blush, most speakers would agree that *novel*, *biography*, and *textbook* are hyponyms of *book*; but so are *paperback* and *hardback*. Now, prototypically, the co-hyponyms of a hyperonym are incompatibles, like *dog* and *cat* of *animal*, and *apple* and *orange* of *fruit*. The hyponyms of *book*, on the other hand, fall into two sets, which display within-set incompatibility, but between-set compatibility. So, for instance, if something is a novel, then it is not a textbook, and if it is a paperback, it is not a hardback; however, there is no embargo on something being simultaneously a paperback and a novel. There are grounds, therefore, for saying that the facets of *book* behave as separate relational entities: *Novel* (or perhaps the [TEXT] facet of *novel*) is a hyponym of the [TEXT] facet of *book*, and *paperback* (or its relevant facet) is a hyponym of the [TOME] facet of *book*. Notice that *novel* gives no information as to the physical format, and *paperback* gives no indication with regard to content.

Another potentially relevant unit for relational purposes is the *microsense* (called *subsense* in Cruse, 1995). A good example of a lexical item exhibiting microsenses is *knife*, which is treated by virtually all dictionaries as a single sense, but can be classified in different domains under a variety of hyperonyms, and under each hyperonym, the sister incompatibles are also different: *cutlery, weapon, surgical instrument, tool*, etc. Microsenses typically show other symptoms of autonomy besides relational ones, including truth-conditional independence. They differ from prototypical senses, however, in that they can be subsumed under a global reading: *knives of various sorts*. (Notice that this is different from the global reading of *book*, which is not a hyperonym in the sense of an abstraction of the features common to the two specific readings, but a union of the two facets.)

It is not clear how far we can validly go in subdividing senses in a search for the ultimate relational atoms. An extreme solution would be to relativize completely to context and say that the proper unit for lexical relations is word-in-context. An undesirable consequence of this solution would seem to be that it would largely destroy any notion of structure in the vocabulary, and would lose sight of the evidence of the 'crystallization' of autonomous units of sense. On the other hand, autonomy appears to be a continuously graded scale, ranging from full polysemous or homonymous senses to pure contextual modulation (like the "male cousin" reading of *cousin* in *My cousin married an actress*), and the setting of a qualifying degree of autonomy for relational units would seem to involve an unavoidable element of arbitrariness.

2.2 Lexical or Conceptual Units?

There is a quite distinct question concerning the 'relata' of sense relations (which, however, interacts with the previous one), namely: Are they truly lexical (i.e., linguistic) relations, or are they in the last analysis conceptual relations? (Notice that the question 'What is a conceptual unit?' is no clearer than 'What is a unit of sense?') Obviously the answer given to this question is going to be heavily theory-dependent; I propose to examine the question from a cognitive linguistic point of view, that is to say, I shall take it for granted that meaning is fundamentally conceptual in nature. This does not, however, render the question otiose.

There seems no reason to doubt that, for instance, the concept CAT stands in a hyponymous (or hyponym-like) relation to the concept ANIMAL, and this could be a satisfactory 'explanation' of why *cat* is a hyponym of *animal*. But when informants judge the words *cat* and *animal*, are they making a direct conceptual judgement, or are they making a lexical judgment? The fact that informants (at least my student informants) judge *cat* to be a better hyponym of *animal* than *pussy* or *moggy*, suggests that they are judging the words. However, this conclusion depends on another assumption, namely, that *cat, pussy,* and *moggy* map onto the same concept. Unfortunately, there are at present no clear criteria for individuating concepts. Let us provisionally adopt a truth-conditional criterion: A conceptual difference entails and is entailed by a truth-conditional difference. On this assumption, *cat* and *pussy* map onto the same concept, and so informants must be judging lexical items. But in that case, we have the problem of 'locating' the difference between *cat* and *pussy*: If it is not in the conceptual system, where is it? In Cruse (1990) it was suggested that lexical items might have meaning properties not accounted for by their associated concepts. However, it is not really clear what this means. Perhaps words like *pussy* have in addition to a concept-mapping (in this case, identical to that of *cat*) some parallel connection with certain non-conceptual areas of the cognitive system, which have the effect of adding expressive or other 'coloring' to the lexical item.

3. DEFINING HYPONYMY

There are various ways of characterizing hyponymy, but a choice of defining strategy will depend on precisely what is being targeted: Are we aiming at an idealized logical notion, or are we trying to characterize speakers' intuitions? This section looks at more idealized approaches.

3.1 Logical Definitions

Taking a logical approach, we can define hyponymy either extensionally or intensionally. One extensional definition is the following, after Cann (1993), but modified to exclude synonymy:

> X is a hyponym of Y iff there exists a meaning postulate relating X′ and Y′ of the form: $\forall x [X'(x) \rightarrow Y'(x)]$, but none of the form: $\forall x [Y'(x) \rightarrow X'(x)]$.

(Here, X′ and Y′ are the logical constants corresponding to the lexical items X and Y, and the definition states, effectively, that for X to be a hyponym of Y, the extension of X′ must be included in the extension of Y′.)

The problem with this sort of definition, at least from the point of view of lexical semantics, is that the relationship is not shown to arise from the meanings (in the ordinary sense) of X and Y; that is to say, we do not know what it is about the meanings of X and Y that gives rise to the relation.

An example of an intensional definition is the following:

> X is a hyponym of Y iff F(X) entails, but is not entailed by F(Y).

(This sort of definition, but not in this exact form, was first put forward by Lyons (1963). Here, F(-) is a sentential function satisfied by X and Y.)

The propositional relation of entailment is closer to lexical semantic concerns than the relation of material implication that appears in the extensional definition above, as it is understood as arising directly from meaning. But the intensional definition also has serious problems, because although it works in certain obvious cases (*It's a dog* unilaterally entails *It's an animal*), it is not generally the case that F(X) entails F(Y). There are cases where F(Y) entails F(X):

It's not an animal entails *It's not a dog.*
All animals breathe entails *All dogs breathe.*
If it's an animal, then it breathes entails *If it's a dog, then it breathes.*

These are comparatively easily dealt with using ordinary logical notions; more difficult are cases where there is no entailment in either direction:

John became a Methodist./John became a Christian.
Mary was disappointed that John gave her a rose./Mary was disappointed that John gave her a flower.

and cases where there is entailment but no hyponymy:

The boil is on Mary's elbow entails *The boil is on Mary's arm.*

As far as I can see, an extensional definition will encounter the same problems. An ideal solution would enable us to predict all entailment relations resulting from replacing a hyponym with its hyperonym and would enable us to discount, in a principled way, entailments that do not arise from hyponymy. To the best of my knowledge, no such solution has been put forward.

3.2 Collocational Definitions

A rather different approach to the characterization of hyponymy utilizes the notion of collocational normality. It seems reasonable to suppose that the more specific a word's meaning, other things being equal, the more restrictive will be its requirements for collocational normality. It might therefore be suggested that a definition of hyponymy could be framed along the following lines:

X is a hyponym of Y iff the normal contexts of X are a proper subset of the normal contexts of Y.

Another way of phrasing the definition, which recognizes that normality is a graded phenomenon is:

X is a hyponym of Y iff there are contexts where Y is more normal than X, but none where X is more normal than Y.

Straightforward examples of the relative restrictiveness of specific items are:

*The dog purred./The animal purred.
*The dog laid another egg./The creature laid another egg.
*The oak produces pears every other year./The tree produces pears every other year.
*This jacket is worn on the head./This garment is worn on the head.

There is no doubt that this definition embodies a significant truth concerning the contextual relations of hyponyms and hyperonyms, but as with the logical definitions (which also embody important truths), the application to ordinary language is beset with problems.

One problem is that there are many context types where the relationship described above does not hold, that is to say, contexts where the hyperonym is the less normal. In principle, such cases should be describable, and capable of being explicitly discounted in the definition, but it is at present not clear how this should be done in general terms. The following are some examples. First, there are contexts where hyponym and hyperonym occur together:

dogs and other animals   *animals and other animals
*creatures and other animals

Second, in certain contexts, a specific term is needed to avoid pleonasm:

My aunt is an actress.   *My aunt is a woman.
He bit it with his incisors.   *He bit it with his teeth.
It was wine that Mary drank.   *It was a liquid that Mary drank.

Third, there are cases where the context 'engages' principally with the portion of meaning that distinguishes the hyponym from its hyperonym:

a virtuoso pianist   ?a virtuoso human being
an out-of-tune piano   ?an out-of-tune object

A different problem is that if we have a pair of propositional synonyms, one of which is marked either for style or field, while the other is unmarked, the former will come out, by the definition, as a hyponym of the latter. For example, the normal contexts of both *kick the bucket* and *pass away* are subsets of the normal contexts of *die*, and the same is true of *pussy* and *cat*. In a sense, this is a 'correct' result, since the marked synonyms have in some sense 'more meaning' than their unmarked sisters, and it is this extra meaning which restricts their collocational normality. We need therefore to consider whether the usual restriction of the term *hyponymy* to cases where there is a propositional difference between hyponym and hyperonym is justified. There are certainly good reasons for according a special status to propositional hyponymy, even if the term *hyponymy* is used to cover the non-propositional variety. Prototypically, a set of hyponyms under a hyperonym serves to articulate the domain designated by the latter; that is their function, and synonyms do not do this. Also, it is only propositional hyponymy that predicts the well-formedness of expressions formed on the pattern of *X's and other Y's* (and many similar):

dogs and other animals
*animals and other dogs
*dogs and other cats
*doggies and other dogs

3.3 Componential Definitions

A disadvantage of all the foregoing definitions of hyponymy is that although they are, to varying degrees, meaning-dependent, they are not explanatory, in that they do not specify what it is about the meanings of two lexical items that makes one a hyponym of the other.

The answer to this might seem relatively straightforward if one adopts a strict componential approach, in which semantic features are both necessary and sufficient. In such a system, X is a hyponym of Y iff the features defining Y are a proper subset of the features defining X. (This is fully explanatory, of course, only if the feature specifications are established independently of the relations they are used to define.) By this criterion, *stallion* ([MALE][EQUINE][ANIMAL]) is unproblematically a hyponym of *horse* ([EQUINE][ANIMAL]).

The componential definition given above is of course only applicable to cases where the meanings of both words are characterizable in terms of necessary and sufficient features. But, to take a famous example, although *chess* is uncontroversially a hyponym of *game*, if Wittgenstein is correct, it cannot be shown to satisfy the definition, because *game* cannot be defined in terms of necessary and sufficient features (I leave it open whether *chess* can be so defined).

4. TOWARDS A PROTOTYPE-THEORETICAL CHARACTERIZATION OF HYPONYMY

A criticism that could be leveled at all the definitions that we have considered so far is that, even if they could be made to work, they are too strict: They do not characterize native speaker intuitions concerning the relationship. For instance, informants asked to pick out pairs from a list which exemplify the same relationship as *tulip:flower* and *apple:fruit* unhesitatingly assign that status to *dog:pet*, even though *It's a dog* does not entail *It's a pet*. Now, it was stated above that propositional hyponymy is crucial to the well-formedness of expressions such as *X's and other Y's*. However, this is not strictly true if hyponymy is defined logically, since *dogs and other pets* is perfectly well-formed. Facts such as these suggest that there is a 'natural category' of hyponymy (or hyponymous pairs), which is a valid object of inquiry in its own right. One well-established approach to the description of natural categories is prototype theory, and it seems worthwhile exploring the possibility of characterizing hyponymy in this way.

4.1 A Prototype Feature Characterization of the Relationship

A standard way of characterizing a conceptual category within prototype theory is in terms of a set of features, not unlike a classical componential analysis. These features are not, however, individually necessary and jointly sufficient (although there is no embargo on individual features being necessary or sufficient), but they do have the property that the more of them an item (or sub-category) possesses, the better an example it is of the relevant (hyperonymic) category. In more sophisticated versions, features are differentially weighted for 'importance'. In principle any list of features (and their weightings) can be empirically justified in terms of speakers' goodness-of-exemplar (GOE) ratings. Basically, the aim is to specify features that a pair of lexical items must show in order to be judged good examples of hyponymy. A first attempt was made in Cruse (1994), but this had a number of drawbacks. The following, still only tentative, is a revised attempt.

Features determining the GOE score for X as a hyponym of Y:

1. There is no inherent categorial incongruity between X and Y.

This means that X and Y must be (literally) predicable of the same argument without anomaly. This feature can be regarded as 'necessary'.

2.   The truth of *A is X* leads to an expectation of the consequent truth of *A is Y* that is greater than the reverse expectation.

Other things being equal, the greater the expectation from X to Y, the greater the GOE score. (Here 'expectation' must be understood as including 'logical necessity' as a maximum value.) It may be necessary to specify a minimum value.

3.   Expressions of the form *An X is a kind/sort/type/variety of Y* are normal.

This feature is intended to capture the intuition that, for instance, *cat:animal* better exemplifies hyponymy than, say, *kitten:cat*.

4.   There is no lexical item Z such that Z is a hyponym of Y and a hyperonym of X.

This feature is intended to capture the intuition that *spaniel:dog* and *dog:animal* are better examples of hyponymy than *spaniel:animal*.

5.   X and Y are matched in respect of their non-propositional features (such as expressiveness, register affiliation, etc.)

This feature captures the intuition that *cat:animal* is a better example of hyponymy than *pussy-cat:animal* or *moggy:animal*.

Features 1-4 are arguably primarily conceptual and only secondarily lexical in nature; feature 5, however, could be seen as purely lexical. If the features were weighted, presumably features 1 and 2 would receive the highest weighting.

The prototype definition of hyponymy given above has not been empirically tested, but it does not seem implausible that something on similar lines would turn out to be predictive of, or correlate with, GOE judgments. However, this is the only respect in which it would be superior to a logical definition: Its explanatory value would be no greater. If a prototype-theoretical approach to lexical meaning is adopted, then each term in a putative relationship will have a prototype characterization. It seems reasonable to expect that systematic relations between lexical meanings such as hyponymy should be predictable from the prototype characterizations of the relevant lexical items (as they are from a classical feature analysis). However, attempts to implement this requirement run into a number of problems.

4.2 Predicting the Relationship from Characterizations of the Relata

As soon as we start thinking about how hyponymy (and incompatibility) might be inferred from prototype representations of concepts, one central point becomes immediately clear: It will not be possible, unless category boundaries are somehow represented (or are at least inferable from representations). Most current models do not indicate category boundaries, and in fact this whole topic is one which has suffered a relative neglect in cognitive linguistics. The basic problem arises from the fact that because the features defining a category are not necessary and sufficient (and have different weightings), it is no longer clear how to distinguish marginal members of a category from non-members, a distinction presupposed by the very notion of hyponymy. (In the move from Aristotelian membership criteria to prototype membership criteria we gained the ability to account for the internal structure of categories, and for the fuzziness of boundaries, but we lost the ability to locate the boundaries.)

One prototype account of category structure which takes on board the need for boundaries is that of Hampton (1991). Hampton's prototype representation consists of a list of features, each one weighted for 'importance'. To determine whether X is a member of the category C, we first list which of the defining features of C are present in X. Then we multiply each feature by its weighting and add the results together to obtain a 'satisfaction rating' for X. Each category comes with a threshold value of the satisfaction rating, which is the criterion for membership, and effectively sets the boundary to the category. We can use Hampton's approach as follows to diagnose hyponymy.

One suggestion as to how to determine whether C(1) is a hyponym of C(2) involves the steps below. (NB: Hampton claims that for subjects to assent to the statement *An X is a Y*, it is only necessary for the prototype of X to reach the threshold of Y. This simplifies the calculation. As a further simplification, it will be assumed that all the features in the specification of a category are necessary for the prototype.)

- Extract all the features in the specification of C(2) that appear in the specification of C(1).
- Multiply each of the extracted features by its weighting in C(2).
- Add together the scores for all relevant features.
- Compare the result with the 'threshold value' for C(2).
- Carry out the same process with C(1) and C(2) reversed.

If the global 'importance rating' for C(1) reaches the criterial value for C(2), and vice versa, then C(1) and C(2) are some sort of synonyms; if the global 'importance rating' for C(1) reaches the criterial value for C(2), but the score for C(2) does not reach the criterion for C(1), then C(1) is a hyponym of C(2); if neither C(1) nor C(2) reaches the criterion for the other, then C(1) and C(2) are either heteronyms (like *father* and *architect*) or incompatibles (like *cat* and *dog*). This can be seen as a possible prototype equivalent of the method given earlier for determining hyponymy from a classical feature analysis.

This method of characterizing hyponymy has a number of drawbacks, and it is not at all clear how they might be remedied. For instance, category membership is portrayed purely as a matter of the possession or non-possession of (a sufficient number of) 'positive' features. However, it is clear that category membership is at least as much a matter of the non-possession of 'negative' features. This point is dramatically highlighted if one tries to apply the method to the diagnosis of incompatibility: There is, in fact, no way of distinguishing 'mere heteronyms' like *father* and *architect*, from incompatibles like *cat* and *dog*. This is an important distinction, but unfortunately it is not possible to decide between the two on the basis of the sort of prototype representation that we have up to now been dealing with. Note that in order to prove incompatibility, we have to show that if something reaches the criterion for, say, *dog*, it will automatically fail to reach the criterion for *cat*. This is logically impossible unless we incorporate into prototype representations features with negative weighting. (Thus, for instance, "has soft fur", "purrs when stroked", "moves by hopping", "larger than average human", and "has scales" would all have negative weighting for *dog*.) This is effectively the prototype equivalent of Katz and Fodor's 'antonymous n-tuples'. To the best of my knowledge, no proposals within prototype theory have taken this requirement on board. Negative features are not only relevant to incompatibility: Incompatibility requires a degree of negativity above some criterial level; for hyponymy, although some negative features are tolerated (in the case of marginal subcategories), they must not exceed some criterial level.

As it stands, the algorithm suggested above for determining hyponymy gives a yes/no answer, rather than a graded answer. There are various possible ways of getting a 'degree of hyponymity' rating. One is to say that the closer the satisfaction rating for C(1) approaches the criterion for C(2) membership, the better the GOE rating of C(1) as a hyponym of C(2). Another possibility is to say that the greater the proportion of items reaching the criterion for C(1) (other than the prototype) that also reach the criterion for C(2), the better the GOE rating of C(1) as a hyponym of C(2).

Several problems arise from the fact that category boundaries are represented by a single global feature count. First of all, there is little or no evidence that category boundaries can be so represented. Secondly, the calculation of a graded hyponymity rating is considerably more difficult if the criterion for a category can be reached by different combinations of features. Setting limits on individual features might be an improvement, but even that would discount any interaction between features, and it seems unlikely that this does not occur. But this does not exhaust the problems: The definition of hyponymy, as well as the definition of incompatibility, must not only specify some sort of criterial value for 'positive' features, but also a maximum level for 'negative' features (with, in all probability, some sort of trade-off relations between the two).

It should be clear from the discussion in this and the previous section that a fully satisfactory characterization of hyponymy has not yet been achieved.

5. VARIETIES OF HYPONYMY

Up to now, the discussion of hyponymy has centered on a generic notion of the relation. However, there are significant sub-varieties which deserve attention.

5.1 Taxonymy

Lyons states that taxonomic lexical hierarchies are structured by the relations of hyponymy and incompatibility. This is true as far as it goes, but it is necessary to make a distinction between two relations of inclusion. The first is the relation which is exemplified in *An X is a Y* (which corresponds to 'simple' hyponymy); the second is the relation for which *An X is a kind/type of Y* is diagnostic, which is more discriminating than hyponymy, and which functions as the 'vertical' relation in a taxonomy. In Cruse (1986), the second relation is called taxonymy. (Each of these relations of inclusion has a partner relation of exclusion: *A Z is not a Y* [incompatibility] and *A Z is a different kind/type of Y than X* [co-taxonymy].) It is perhaps noteworthy that Wierzbicka (1996) includes "a kind of" as one of her semantic primitives, and does not subject it to further analysis; here, an attempt is made to elucidate its nature.

Taxonomizing is not simply a matter of dividing a larger class into smaller classes. Some logically impeccable subdivisions do not yield good taxonomies:

?A stallion/mare/foal is a kind/type of horse.
(A stallion is a horse.)
?A blonde/queen/actress is a kind of woman.
(An actress is a woman.)
?A red/green/blue hat is a kind of hat.
(A red hat is a hat.)
A mustang is a kind of horse.
An ash-blonde is a kind of blonde.
A Stetson is a kind of hat.

It is obvious that the expression *is a kind/type of* exerts some kind of selectional pressure on pairs of items. (There are indications that *X-ing is a way of Y-ing* has a similar selectional effect on verbs; the relation does not seem to be relevant at all to adjectives.) There are various possibilities here. One not very attractive possibility is that pairs of lexical items are simply marked for their ability to enter the *is a kind of* construction, and we simply have to learn this alongside (or as part of) their meanings. For instance, *mustang* could have as part of its semantic specification a feature like [KIND OF HORSE], whereas *stallion* would not possess this feature, and would therefore not satisfy the selectional requirements for '*is a kind of*'. (Alternatively, the lexical entry for *horse* could include a list of its possible kinds.) A more interesting possibility is that *is a kind of* selects for a particular relationship between hyperonym and hyponym, which can be described independently of particular pairs of items and which can be inferred from representations of categories concerned. What we seem to be looking for is a 'principle of taxonomic subdivision'.

5.1.1 Taxonymy and the Relations Between Sub-classes

One approach to the characterization of taxonomy is to think of the nature of the resultant sub-categories and their relations to one another, in the light of the purpose of taxonomization. Taxonomy exists to articulate a domain in the most efficient way. This requires "good" categories, which are (a) internally cohesive, (b) externally distinctive, and (c) maximally informative.

In many of the instances where a good hyponym is not a good taxonym of a hyperonym, there is a straightforward componential analysis of the hyponym in terms of the *hyperonym* plus a single feature, as in:

stallion = [MALE][HORSE]
kitten = [YOUNG][CAT]
(cf. also *blonde, actress*)

and so on. That is to say, many inadmissible taxonyms are nominal kinds; a significant number of good taxonyms, on the other hand, seem to be natural kinds, which typically are not easily definable in terms of their hyperonyms, and require encyclopedic characterization. In Cruse (1986) it was suggested that single-feature category division might be responsible for the fact that *stallion, kitten,* and *blonde* are not satisfactory taxonyms of *horse, cat,* and *woman,* respectively, because it creates non-optimal categories. (For instance, a division of the domain of animals into males and females would yield categories with too little internal cohesion and too little external distinctiveness, in that, for instance, a female mouse resembles a male mouse more than a female elephant.).

However, there are problems with this analysis. It is not difficult to find cases where a satisfactory taxonomy seems to be founded on a single-property division. Take the case of spoons: These are taxonomized on the basis of what they are used in connection with (*teaspoon, coffee-spoon, soup spoon,* etc.). (It is significant that neither *large spoon, metal spoon, round spoon,* nor *deep spoon* is a satisfactory taxonym of *spoon*:

A teaspoon/soup spoon is a type of spoon.
?A large/metal/round/deep spoon is a kind of spoon.)

Perhaps the explanation lies in the nature of the features? Some features are conceptually "simple" (LARGE, ROUND, etc.), others are more "complex" (FUNCTION?): Perhaps complex features are better able to support taxonomy? What, then, about the subdivision of blondes, which relies on shade of hair-color (*An ash-blonde/strawberry blonde is a kind of blonde*)?

The fact is, there are reasons to believe that the problem of taxonomy cannot be solved merely by looking at the nature of the subcategories: Take the domain of BLONDES. *Ash-blonde* and *strawberry blonde* are satisfactory taxonyms of *blonde,* but they are by no means the optimal sub-categories of BLONDE. We can get much 'better' subcategories than ASH BLONDE and STRAWBERRY BLONDE by dividing the domain into BLONDE HOUSEWIVES, BLONDE NURSES, BLONDE WAITRESSES, BLONDE ACTRESSES, etc., but these are not good taxonyms of *blonde.*

Suppose that a particular species of bird has a number of varieties and a very marked difference between males and females. In such a case, it is not inconceivable that a male/female division would yield the best categories, in that the males of different varieties resembled one another more than the male and female of a particular variety. However, even if that were the case, a sex-based division would still be taxonomically 'wrong'.

A given category may be a satisfactory subdivision of one hyperonym, but not of another. For instance, *?A waitress/prostitute is a kind of woman,* is not good, but *A prostitute is a type of sex-worker* is OK. If the crucial factor were the nature of the resultant category, this would be hard to explain. It is possible that the good category principle has a role to play in the characterization of taxonymy, but is subordinate to some other principle or principles.

5.1.2 Taxonymy and 'Perspective'

It seems then that a satisfactory taxonym must engage in a particular way with a particular aspect of the meaning of the hyperonym. In Cruse (1994), it was suggested that taxonym and hyperonym must share the same *perspective.* The reason *stallion* is not a good taxonym of *horse,* it was argued, is that it has a 'sexual' perspective, while *horse* does not; the reason *blonde* is not a good taxonym of *woman* is that it adopts a 'hair-color' perspective, while *woman* does not. *Ash-blonde,* on the other hand, has the same 'hair-color' perspective as *blonde,* and that is why it is a satisfactory taxonym.

A problem with this proposal is that the notion of perspective is somewhat vague and needs to be made more precise. This is not easy: Two possible lines of approach may be singled out. The first possibility involves the notion of 'highlighting'. The idea is that in some complex lexical meanings, certain features or conceptual areas are more salient than others. For instance, it seems intuitively clear that *stallion* and *mare* highlight the feature of sex, and *blonde* and *brunette* highlight the feature of hair-color. To put a little more flesh on the notion, we can say that a highlighted feature distinguishes a lexical meaning from other members of its default contrast set, whereas a non-highlighted feature is often presupposed, and indeed shared, with the members of the default contrast set. Thus, the default contrast of *stallion* is *mare,* while HORSE is presupposed by both *stallion* and *mare.* That is to say, the default assumption on hearing *That's not a stallion,* is that the referent is a mare. Similarly, on hearing *X is not a blonde,* the default assumption is that X is a brunette, or a red-head, etc., with WOMAN presupposed.

(Notice that just because a word possesses the feature FEMALE, say, even if it is a necessary feature, it does not follow that that feature is highlighted. For instance, *mother* contains FEMALE as a necessary feature, but it is not obvious that the default contrast of *mother* is *father,* or that PARENT is presupposed. In *X is not my mother,* for instance, the default contrast is intuitively with items like *sister* and *aunt,* and perhaps even also *neighbor* and *friend,* and FEMALE is presupposed.)

Let us assume that we can identify highlighted features in the way proposed above. The proposal in connection with taxonymy is that a taxonym must (a) further specify a highlighted feature of the hyperonym and (b) must similarly highlight it. From this it will follow that the reason *stallion* and *foal* are not good taxonyms of *horse,* is that they highlight features not highlighted in *horse;* a similar explanation holds for *blonde* and *woman;* likewise, *prostitute* is not a good taxonym of *woman* because it highlights profession, but is a good taxonym of *sex-worker,* because it specifies further and highlights what is highlighted in *sex-worker,* namely, type of work.

So far so good. But it is arguable that this account does not go far enough. It can perhaps explain why *stallion* is not a good taxonym of *horse,* but it does not tell us what IS a good taxonym of *horse:* What does *horse* highlight? (The same question could be asked of artefactual kinds such as *chair, spoon* and *violin.*) One answer is that it does not highlight anything, and therefore its taxonyms must likewise not highlight anything. I must confess I am slightly tempted by this, but at the same time I do not find it fully satisfying.

5.1.3 Taxonymy and 'Essence'

A second suggestion for specifying the notion of 'perspective' is more tentative, but may be closer to the truth. It is that a good taxonym must have as its essence a specification of the essence of the hyperonym. Given that the essence of a *blonde* is the possession of fair hair, it follows that the taxonyms of *blonde* must specify fair hair more restrictively; if it is the essence of a *sex-worker* to perform sexual activities for money, then a taxonym must specify those activities further; if the essence of a *spoon* is to fulfill a particular function, then the taxonyms of *spoon* must have as their essence a more specialized function. What, in that case, is the essence of *horse*? This is more difficult, but one possibility is its species; intuitively, *mustang* would seem to have a more highly delimited version of the same essence. (The notion of 'essence' presented here is intended to be an intuitively accessible aspect of meaning [or at least, of category structure], which can be approached through expressions like *the X itself*, or *the real X*, and so on. No philosophical claims are being made.)

A question that is pertinent at this point is whether highlighting and essence are the same thing. One way to try to answer this is to look for cases where they give different predictions regarding taxonyms. There are some indications that they are different, and that it is essence which gives the correct prediction. Take the case of *woman*. It seems reasonable to assume that *woman* highlights [FEMALE], and presupposes [HUMAN], on the grounds that the default contrast of *woman* is *man*. On the assumption that highlighting governs taxonomy, we would predict that taxonyms of *woman* would specify sexuality more narrowly. Hence, we would expect *lesbian* to be a satisfactory taxonym. But this does not seem to be the case. When asked to suggest types or kinds of women, people tend to offer things like *career woman, nest-builder, femme fatale*, and so on. Now these are not subdivisions of sexuality (although some might have sexual consequences), but of personality or character, and it is not implausible (it is even reassuring!) that these come out as the essence of womanhood (a similar result is obtained for *man*). This is, of course, highly tentative, and clearly more solid criteria for determining the essence of something are needed before the argument can be taken any further; but my conviction is growing that it is indeed essence which lies at the heart of the taxonomic endeavor. One further point is worth mentioning: Some essences seem inherently to lend themselves more readily to repeated subdivision than others. Intuitively, SEX, for instance, does not appear to be indefinitely subdivisible, whereas COLOR, FUNCTION, OCCUPATION, and so on are. Perhaps this, too, should be considered one of the prototype features of taxonymy.

The relation of taxonymy presents yet another problem for the enterprise of predicting the sense relations holding between the members of a pair of conceptual categories from prototype representations of those categories. At the very least, prototype representations will have to specify essences, or contain information from which essences can be inferred.

5.2 The Significance of Taxonomic Levels

The exact nature of a hyponymous relation depends to some extent on the taxonomic level of the related items. It is generally agreed that taxonomic hierarchies are centered on the so-called 'basic level' categories. These are informationally rich, highly differentiated categories, which represent the default level of specificity for everyday reference (e.g., *dog* as opposed to *animal* or *spaniel*). They tend to be the first items learned, and represent the highest taxonomic level at which items are associated with clear visual images or patterns of behavioral interaction (one can describe easily enough how one interacts with a dog, but not how one interacts with an animal).

Other taxonomic levels are defined relative to the basic level: Superordinate categories are more inclusive, and subsume more than one basic level category; subordinate categories are included within basic level categories. Superordinate and subordinate categories are usually held to be different in character from basic level categories, and it presumably follows that hyponymous relations also differ according to the levels of the related items. Superordinate categories are typically clearly distinct from their sister categories, but are internally much less unified, in the sense that their members have fewer features in common, than basic level categories. What features they do have in common tend to be abstract, rather than 'concrete' sensory-motor features. They are sometimes described as 'parasitic' on basic level categories, in that (a) the natural way to explain them is to say 'things like cats, dogs, horses, etc.' (for *animals*) or 'things like chairs, tables, etc.' (for *furniture*). Also, it is difficult to visualize a superordinate category except in terms of one or more specific basic level hyponyms. However, it is possible to underestimate the extent of common features in superordinate categories if inappropriate means of extracting them are used. For instance, Rosch & Mervis (1975) attempted to determine the defining features for superordinate categories by examining subjects' lists of features for subsumed basic level categories. For instance, lists of features produced for *chair*, *table*, *bed*, *cupboard*, and so on revealed virtually nothing common to all of them, which led to the notion that a category like FURNITURE was virtually empty of content and should be regarded as a "collective concept . . . which stands for a heterogenous collection of things of different kinds" (Wierzbicka, 1996, p. 155). However, asking for features for *chair*, *table*, *bed*, *cupboard*, and so on, automatically highlights the features which differentiate them and backgrounds the features they have in common: To elicit the distinctive features of, for instance, FURNITURE, one must contrast the category with its sister categories at the same level, such as SOFT FURNISHINGS, APPLIANCES, and FIXTURES (such as shelves, built-in cupboards, and so on). When this is done, it becomes clear that FURNITURE is a more cohesive category than it at first appeared, having as features, "movability", "rigidity", and "locational (rather than agentive or instrumental) function" (cf. Bolinger, 1992; Cruse, 1992).

Subordinate categories are highly cohesive, but less distinctive than basic level categories. Obviously, the features which distinguish a spaniel from a collie are far fewer than those that distinguish a cat from a dog. But it is arguably a mistake to claim, as some do, that subordinate categories are typically single-feature categories (cf. Ungerer & Schmid, 1996), i.e., categories differentiated from their immediate hyperonyms by the presence of a single feature. This seems to be a misinterpretation of the undeniable fact that subordinate categories are frequently designated linguistically by the name of the hyperonymic basic level category plus a distinguishing modifier:

tit(mouse):	blue tit, coal tit, great tit, long-tailed tit
spoon:	teaspoon, soup spoon, dessert spoon
cat:	Manx cat, Persian cat, tabby cat

Although the modifier in these cases highlights a single salient distinguishing feature, it by no means follows that that is the only difference between the subcategories. For instance, while it is true that a long-tailed tit has a longer tail than any of the others, it is also smaller, differently colored, and has different habits and habitat. In other words, the sort of extra specificity shown by long-tailed tit over tit is not different in kind from that of tit over bird. The same point could be made (not quite so convincingly) about the subvarieties of spoon: The labels indicate what they are used in conjunction with, but they also differ in size and shape, and in the way they are used (for instance, a teaspoon is not used for conveying a liquid to the mouth, as a soup spoon is).

5.3 Types of Superordination and Subordination

There are different ways of subdividing a category, irrespective of its hierarchical level, and we can therefore distinguish types, or modes, of hyponymy. Three modes will be distinguished here: the natural kind mode, the nominal kind mode, and the functional kind mode.

5.3.1 The Natural Kind Mode

Natural kinds are things like animal and plant species, such as lion, eagle, trout, and oak, and naturally occurring materials, such as water, soil, rock, iron, and wood. The names of natural kinds behave to some extent like proper names in that they show referential stability in the face of quite radical changes in the speaker's beliefs concerning the referent(s). For instance, suppose one had a friend called John whom one believed to be a man. One day one sees this person naked and is astonished to observe the bodily appurtenances of a woman. One might well continue to refer to the person as 'John', but one would no longer refer to 'him' as a man. Something similar occurs with natural kind terms. If it were discovered one day that the substance everyone called *water*, and which everyone believed to be a compound of hydrogen and oxygen, was actually a compound of krypton and oxygen, people would not say *So it's not water after all*, but rather *Water isn't what we thought it was*. Natural kinds can be described by a set of features, but such descriptions are in a sense provisional: The essence of a natural kind remains mysterious (although we tend to assume that there is an abiding essence).

The relevance of natural kinds to hyponymy is that the relation of a natural kind to a hyperonym cannot be captured by a single feature or small set of features: It is in essence encyclopedic (think of what must be 'added' to ANIMAL, in the way of semantic features, to make DOG). This is not determined by hierarchical level: The specification of TIT relative to BIRD is of the same type (i.e., encyclopedic) as LONG-TAILED TIT relative to TIT. It is arguable that at least some artefactual kinds have similar properties to natural kinds. Lyons (1981) argues that VIOLIN has much the same sort of relation to MUSICAL INSTRUMENT as, say, DOG has to ANIMAL. (It is arguable that there are verbal [e.g., *sing, walk*] and adjectival [e.g., *red, happy*] equivalents of natural kinds.)

5.3.2 The Nominal Kind Mode

In contrast to natural kinds, the relation between what some call nominal kinds and their hyperonyms CAN be captured in terms of a single differentiating feature, as in the case of *mare:horse, kitten:cat, blonde:woman*. In many cases, the relationship parallels that between a morphologically derived term and its base (*lioness:lion, duckling:duck*). Prototypically, a nominal kind term is not a 'kind, sort, or type' of its hyperonym, whereas a natural kind term is. Thus a stallion is not a type of horse (although a mustang is), nor is a kitten a type of cat (see the discussion of taxonymy in Section 5.1 above).

5.3.3 The Functional Mode

Many nouns incorporate functional features in their definitions and it is not infrequently the case that the function, or at least an aspect of it, is captured by the meaning of a hyperonym. This is the case with, for instance, *gun:weapon, hammer:tool, jacket:garment, beer:beverage, car:vehicle, violin:musical instrument*. In these cases the hyponym simultaneously specifies the function more precisely and adds perceptual features which are largely absent from the hyperonym. In many respects these 'functional kinds' share characteristics with natural kinds. For instance, their specification is encyclopedic rather than single-featured: Although "male horse" is a satisfactory definition of *stallion*, there is no X such that "an X weapon" is a satisfactory definition of *gun*, or Y such that "a Y vehicle" is a satisfactory definition of *car*. This parallels the fact that there is no P such that "a P animal" is a definition of *horse*. (A feature such as EQUINE does not decompose "horse", but is parasitic on it, and means "resembling or pertaining to horses".) It is also noteworthy that inherent functional kinds are typically 'types of' their hyperonyms. It is fairly clear what common properties functional hyperonyms 'extract' from their hyponyms; it is less clear, however, what in general terms taxonomic hyperonyms capture.

Words like *gun, car, jacket*, and so on can be described as inherently functional, in that their function is, as it were, inscribed in the fabric of their inherent meaning. But what about cases like *dog* and *pet*? PET is undoubtedly a functional category, but there are no categories of animals that have pet-hood encoded in their meanings. It is therefore perhaps surprising that speakers are so ready to include *dog:pet* as exemplifying hyponymy. Notice that although there are non-inherent functional categories, there are no non-inherent taxonomic categories of the ANIMAL type.

5.4 Paradigmatically Non-congruent Hyponymy

Hyponyms are prototypically of the same syntactic type as their hyperonyms; however, while noun hyponyms universally have noun hyperonyms, there may be a difference of sub-class. For instance, functional hyperonyms of count-noun hyponyms are often mass nouns:

knife/fork/spoon	cutlery
cup/plate/saucer	crockery
envelope/(writing paper)	stationery
jacket/shirt/skirt	clothing
shoe/sandal/boot	footwear
vest/underpants	underwear
sheets/pillow-cases	bed linen

Conversely, mass-noun names of substances frequently have count-noun hyperonyms (these may or may not be functional):

gold/iron/copper	metals
diamond/coal/fluorspar	minerals
beer/wine/coke	beverages
ginger/allspice/pepper	spices
oxygen/hydrogen/nitrogen	gases/elements

6. CONCLUSIONS

Hyponymy is one of the most important structuring relations in the vocabulary. It occurs in a wide range of content domains and in all major syntactic categories, and is intimately implicated in the way languages articulate the world of experience. Although intuitively a fairly straightforward relation of inclusion, easily recognized by non-linguists, attempts to characterize it and its varieties precisely reveal a number of interesting problems. There is some uncertainty, for instance, concerning the nature of the units related by hyponymy: Are they lexical or conceptual? How should they be individuated? There are various ways of approaching the definition of the relation, but many arguably do not capture what it is that speakers are judging when they pick out hyponymous pairs from a list. Prototype-theoretical definitions, which might be expected to be well-adapted to this task, turn out to have quite serious problems. There are several varieties of hyponymy. Hyponymy is particularly prevalent amongst nouns, as are well-developed taxonomic hierarchies, and the way hyponymy manifests itself shows some interesting correlations with the taxonomic level (basic, superordinate, or subordinate level) of the related items. Another dimension of variation is whether the extra specificity of hyponym over hyperonym can be captured by a single feature or whether it requires an encyclopedic characterization. One of the most important varieties of hyponymy (but also one of the most difficult to elucidate) is taxonymy, the relation which determines the well formedness of expressions of the form 'An X is a kind of Y', and is the vertical structuring relation of taxonomic lexical hierarchies.

References

Bolinger, D. (1992). About furniture and birds. *Cognitive Linguistics*, 3, 111-118.

Cann, R. (1993). *Formal Semantics*. Cambridge: Cambridge University Press.

Cruse, D. A. (1986). *Lexical Semantics*. Cambridge: Cambridge University Press.

Cruse, D. A. (1990). Prototype theory and lexical semantics. In S. L. Tsohatzidis (Ed.), *Meanings and Prototypes: Studies in Linguistic Categorization*. London: Routledge.

Cruse, D. A. (1992). Cognitive linguistics and word meaning: Taylor on linguistic categorization. *Journal of Linguistics*, 28, 165-183.

Cruse, D. A. (1994). Prototype theory and lexical relations. *Rivista di Linguistica*, 6, 167-188.

Cruse, D. A. (1995). Polysemy and related phenomena from a cognitive linguistic viewpoint. In P. Saint-Dizier & E. Viegas (Eds.), *Computational Lexical Semantics*. Cambridge: Cambridge University Press.

Hampton, J. A. (1991). The combination of prototype concepts. In P. J. Schwanenflugel (Ed.), *The Psychology of Word Meanings*. Hillsdale, NJ: Erlbaum.

Lyons, J. (1963). *Structural Semantics*. Cambridge: Cambridge University Press.

Lyons, J. (1968). *Introduction to Theoretical Linguistics*. Cambridge: Cambridge University Press.

Lyons, J. (1981). *Language, Meaning and Context*. London: Fontana.

Rosch, E., & Mervis, C. B. (1975). Family resemblances: Studies in the internal structure of categories. *Cognitive Psychology, 7*, 573-605.

Ungerer, F., & Schmid, H.-J. (1996). *An Introduction to Cognitive Linguistics*. London and New York: Longmans.

Wierzbicka, A. (1996). *Semantics: Primes and Universals*. Oxford: Oxford University Press.

Chapter 2

On the Semantics of Troponymy

Christiane Fellbaum
Cognitive Science Laboratory, Princeton University, Princeton, NJ, USA

Abstract:
   The principal relation linking verbs in a semantic network is the manner relation (or "troponymy"). We examine the nature of troponymy across different semantic domains and verb classes in an attempt to arrive at a more subtle understanding of this intuitive relation. Troponymy is not a semantically homogeneous relation; rather, it is polysemous and encompasses distinct sub-relations. We identify and discuss Manner, Function, and Result.
   Furthermore, different kinds of troponyms differ from their semantically less elaborated superordinates in their syntactic behavior. In some cases, troponyms exhibit a wider range of syntactic alternations; in other cases, the troponyms are more restricted in their argument-projecting properties.[1]

1. INTRODUCTION: RELATIONS IN THE LEXICON

   The lexicon contains all those concepts to which speakers of a language attach a label (a word). The labeling process is not arbitrary; we pick out concepts that are important or stand out in some way, such as for their perceptual saliency. If one examines the lexicalized concepts in relation to one another, it becomes clear that they differ in systematic ways that are characterizable in terms of similarities or contrasts. These consistent differentiations among concepts are what we call *semantic relations.*

   Relations are very real, though speakers may be unaware of them and may be unable to articulate them (as is the case with most metalinguistic knowledge). But there are situations when one must consciously confront semantic relations. Building a lexical resource presents one such situation.

2. SEMANTIC RELATIONS IN LEXICONS AND THESAURI

   The structure of a lexical entry in a dictionary reflects the relatedness of words and concepts: The target word is usually defined in terms of a related word and some differentiae. Frequently, the related word is a superordinate, or more general concept; for example, *dingo* is defined in the American Heritage Dictionary as a *wild dog of Australia.* The super-/subordinate relation, or hyponymy (or hyperonymy or ISA) relation works well to characterize the meaning of nouns, as does meronymy, the part-whole relation; thus, the American Heritage Dictionary defines *top* as *the uppermost part, point, surface or end of* something.

Defining meanings in terms of such relations reflects the paradigmatic organization of the lexicon. Many dictionaries also supply syntagmatic relations between the target and other words by means of illustrative sentences. Syntagmatic relations constrain the contexts in which a word may be used and can be seen as a complementary way of representing speakers' lexical knowledge. A third type of knowledge, world or encyclopedic knowledge, cannot be expressed in terms of relations, as it lists information about the concept behind the word in language not bound to formulas such as definiendum-definiens.

A thesaurus lists words in semantically related groups. It is intended for users who have a certain concept in mind, and are looking either for alternative words to express this concept or for words that express similar concepts. Because its purpose is to suggest words that may be substitutable for each other, a thesaurus is necessarily organized paradigmatically. But the semantic relations between the members of a word group are not made explicit, nor are all words within a group related in the same way.

The lexical database WordNet (G. A. Miller, 1990; Fellbaum, 1998) resembles a thesaurus in that it represents word meanings primarily in terms of conceptual-semantic and lexical relations. Relations among groups of cognitively synonymous words are given straightforwardly, without being woven into the definitions, as in conventional lexicography. But unlike a standard thesaurus, the relations are transparent and explicitly labeled; moreover, they have been deliberately limited in number. The resultant structure is a large semantic network for nouns, verbs, adjectives, and adverbs.

The bulk of WordNet, as indeed of any lexicon of English, is comprised of nouns; there are far more distinct noun forms than verbs or adjectives in the language. In constructing a semantic net of nouns, hyponymy and meronymy relations could be applied in a straightforward manner (but see G. A. Miller, 1998, for details). By contrast, adjectives in many cases denote values of attributes, and can be interrelated via antonymy, as in the case of the pairs *hot-cold* and *long-short,* where the antonyms express values of *heat* and *length,* respectively. Most adjectives, like *icy* and *elongated,* have no salient antonyms; they are linked to core adjectives like *cold* and *long* through a relation of semantic similarity (K. J. Miller, 1998).

Unlike nouns and adjectives, verbs do not seem obviously related in a clearly discernible and consistent manner. The complex semantics of verb relations, in particular the one we dubbed "troponymy", will be the focus of this chapter.

2.1 Troponymy

Verbs in WordNet cluster into semantically related groups, such as the verbs of motion, communication, and contact. The members of each such group, which tend to share some syntactic properties and selectional restrictions, express various elaborations of one core, general concept (such as *move, communicate, touch).* The relation of verbs like *run, ride,* and *fly* to the core verb *move,* or of *hit, scrub,* and *kick* to *touch* therefore appears at first blush to be just a variation of the hyponymy relation among nouns: The subordinate concept contains the superordinate, but adds some additional semantic specification of its own.

Despite the apparent similarity, the hierarchical relation among verbs differs from the ISA relation among nouns. First, the formula linking hierarchically related nouns, *X is a kind of Y* sounds odd when applied to verbs: *(To) yodel is a kind of (to) sing;* only when changing the verbs into nouns (gerund) can the formula apply: *Yodeling is a kind of singing.* Second, in the case of nouns, *kind of,* which makes the hierarchical relation explicit, can be omitted from the formula: The statements *A donkey is an animal* and *An orchid is a flower* are equivalent to *A donkey is a kind of animal* and *An orchid is a kind of flower.* By contrast, statements like *Sketching is drawing/To sketch is to draw* or *Murmuring is talking/To murmur is to talk* are odd; here the superordinate is not accompanied by some qualification and the relation between the two activities is not expressed or implied. This indicates that there is more than just an ISA-like relation among concepts expressed by verbs with varying degrees of semantic elaboration, and that the semantic distinction between two verbs is different from the features that distinguish two nouns in a hyponymic relation. The relation among verbs like *murmur* and *talk* may be much greater if it has to be articulated explicitly. The fact that we can omit the phrase *kind of* from the ISA definition points to a greater similarity among hierarchically related nouns than among similarly related verbs.

The ISA formula works best between closely related nouns, and less well between nouns that are far apart in the hierarchy: *A donkey is an animal* is better than *A dog is a (kind of) creature* or *A dalmatian is a (kind of) animal.* But among verbs, the number of levels does not affect the requirement for an explicit modification of the superordinate: *Murmuring is talking/To murmur is to talk* is just as strange as *Murmuring is communicating/To murmur is to communicate.* Clearly, the hierarchical relations among nouns and verbs differ.

To distinguish this apparent hyponymy relation among verbs from the superordinate relation among nouns, we have called it *troponymy* (Fellbaum & Miller, 1991). Rather than expressing "kind", troponymy, which links verbs like *run, ride, fly* to the core verb *move,* or *hit, scrub, kick* to the basic contact verb *touch,* seems to express a manner elaboration. The formula capturing this relation, *to run/ride/fly is to move in some manner* appears to be appropriate.

Like the ISA relation among nouns, troponymy builds hierarchical structures, or "trees", with the semantically most inclusive verb at the root and increasingly specified verbs forming the branches and leaves. While noun hierarchies can be very deep (15 levels or more), verb hierarchies tend to be flatter and "bushy" rather than tall and tree-like. Most verb hierarchies do not exceed 3 or 4 levels.

The manner relation worked well for constructing a large semantic network of verbs (Fellbaum, 1990, 1998). But it became clear that it is in fact highly polysemous, containing itself a number of distinct manner relations. Closer examination of the verbs in hierarchies across different semantic fields reveals that lexicalization involves distinct kinds of semantic elaborations. For example, motion verbs differ in their core meaning along the dimension of speed *(walk vs. run)* or the means of transportation *(truck, bike, train, bus).* Verbs of impact, such as the troponyms of *hit,* may be differentiated along the semantic dimension expressing the degree of force used by the agent *(chop, slam, whack, swat, rap, tap, peck,* etc.). Among verbs of competition, many troponyms are conflations of the root verb *fight* with nouns denoting the occasion for, or form of, the fight: *battle, war, tourney, joust, duel,* etc. Troponyms of communication verbs lexicalize the communicator's intention or purpose *(examine, confess, preach)*, or the medium of communication *(fax, mail, phone, wire)*. Subsets of particular kinds of manners tend to cluster within a given semantic field, where the semantic feature is part of most verbs' semantic make-up, but is present to a varying degree in the related verbs.

Despite its polysemy, troponymy appears to have psychological reality for speakers, who generally accept the *manner* formula for relating troponyms to their superordinates.

3. PSYCHOLINGUISTIC VALIDATION

Fellbaum and Chaffin (1990) performed a number of experiments designed to test the validity of the troponymy relation in WordNet. Despite the great variety of semantic relations that are lumped together under the label troponymy, they found that this relation appears to be meaningful. In the first task, native speakers had no trouble, and showed high agreement among themselves, when they were given verb pairs related via troponymy and were asked to articulate the relation in their own words. While the sentences supplied by the participants differed slightly from each other, they were all clearly variations of the "manner-of" or "way-of" formula.

Second, speakers showed a high rate of agreement with the experimenters when they were asked to sort verbs into pairs that the experimenters had independently labeled as being troponymically related.

In a restricted association task, the participants were given verbs and asked to respond with the first verb that came to mind. The overwhelming number of responses were verbs that were troponymically related to the stimulus verb.

Finally, speakers were presented with an analogy task. They were given pairs of verbs related in various ways, including some troponymically related pairs. Then, presented with single verbs, the participants were asked to identify verbs that, together with the first verb, would make pairs that were related in the same way as the example pairs. The ease with which the participants made up troponym pairs as compared with pairs related in other ways was striking.

The results of these experiments can be interpreted to demonstrate clearly that troponymy is a salient relation, despite its lack of semantic uniformity.

4. DISTINGUISHING MANNERS

Lexical semanticists often take the existence of manner as a component of very many verb meanings for granted. But it is commonly assumed to be a primitive and, to our knowledge, has not been further analyzed. In the remainder of this chapter, we examine a few areas of the verb lexicon where the encoding of a manner component has interesting consequences, and we attempt to draw some distinctions among different kinds of troponymy.

4.1 Motion Verbs

Many motion verbs are regularly decomposed into elements, including one labeled MANNER. Talmy (1985) showed that, for verbs of motion, virtually all human languages follow one of two patterns in their syntax-semantics mapping. One type (including English and Chinese) conflates into a single lexical item the "fact of motion" and a manner component and expresses the path of the motion in a separate adjunct:

(1)   She ran into the room.

The other type of language (including Romance, Greek, and Semitic) conflate the "fact of motion" with a path element and refer to the manner of motion in an adjunct:

(2)   She entered the room dancing.

Such analyses seem to point to the fundamentality of MANNER.

Papafragou, Massey, and Gleitman (2000) compared the performance of English and Greek-speaking children and adults in non-linguistic (memory and categorization) tasks involving motion events and in their linguistic descriptions of these same motion events. The two linguistic groups differed significantly in terms of their encoding, as would be predicted by the Talmy typology (English conflates "fact of motion" and manner, while Greek conflates "fact of motion" and path). But Papafragou et al. found that the performance of the two language groups in the non-linguistic tasks was identical. Given a picture of a man running up a flight of stairs, both groups focused on the path (e.g., "go up") as the main characteristic of the event.

Papafragou et al.'s findings seem to indicate that path is more important as a semantic element than manner; we believe that this must indeed be the case. Imagine events where someone is coming into or leaving a room, crossing a street, or going up or down a flight of stairs, in what is usually called a manner: crawling, limping, slouching, ambling, etc. To refer to the main characteristic of such events, speakers would use sentences like (3). Sentences like (4), which omit the path, would be misleading and violate the maxim for communicating relevant information:

(3)   He came in/came down/went up/went across...
(4)   He crawled/limped/slouched...

(4) would be an appropriate statement only if the moving entity does not proceed along a path or into a direction that can be linguistically encoded, i.e., if the event were an activity with no explicit end state rather than an accomplishment delimited by the path.

The evidence for the greater salience of the path element in the domain of motion verbs suggests that manner is the "semantic icing on the cake".

4.2 Manner vs. Function

Pustejovsky (1995) discusses the semantics of nouns like *pet,* which he calls *roles.* The case of *pet* shows the kind-of relation that holds among so many nouns is not appropriate for role nouns: While one may say that a poodle is a kind of dog, and a dog is a kind of animal, it is distinctly odd to state that *a pet is a kind of dog/animal,* though it clearly is an animal. The relation between *dog* and *animal* differs from that between *pet* and *animal,* and the *kind-of* formula does not apply to the latter pair, as it does to the former. "Pet" is the role that an animal plays in a certain setting that may be culturally dependent. The pet role of an animal may be transient: The pet may be set free and become a wild animal. By contrast, the ISA relation linking *dog* to its superordinate concept *animal* is stable and makes up an inherent part of the meaning of *dog* and *animal; dog* could be cited to illustrate the meaning of animal and vice versa. Other examples of role nouns cited by Pustejovsky are *customer, laundry,* and *groceries,* which refer to functions assumed by concepts only in specific circumstances that can be temporally defined and limited.

A parallel phenomenon seems to exist in the verb lexicon. Many verbs may have, in addition to their strictly denotational meaning, meaning aspects that depend on the context. For example, *run, walk,* and *swim* are manner of motion verbs and *bike* is a verb of motion using a vehicle. But they all are also manners of *exercising,* along with verbs like *row* and *ski.* Both formulae, *to walk/run/swim is to move in some manner* and *to walk/run/swim is to exercise in some manner,* seem appropriate.

To represent these verbs in a semantic network like WordNet, there seem to be two options. Either one interprets a verb like *run* as polysemous, with a sense that is a troponym of *move,* and one that is a troponym of *exercise;* each belongs to a distinct hierarchy. In this case, the semantic similarity between the two senses is lost in the representation. The alternative is to interpret such verbs as monosemous, but to attach them to two distinct superordinates *(move* and *exercise).* But this would entail "tangled" or "intersective" (Cruse, 1986) hierarchies, and make for a potentially very messy net.

Even though the relation of *run, walk,* etc. to *exercise* is represented as troponymy in WordNet, it is clearly not the same as that between *run, walk,* and *move.* A running/walking/swimming event is an exercising event only in certain situations, whereas it is always a motion event. The context-dependent and somewhat arbitrary nature of the concept of "exercise" is reflected in the fact that doctors can simply define certain activities as exercise, such as climbing stairs. *Exercise* is a specific function that verbs like *run, walk,* etc. can take on, in addition to being movements, just as dogs can play the role of pets, in addition to always being animals. We call the relation between *run, walk, dance, bike,* etc. the "function" relation.

The distinction between genuine troponymy and the function relation is reflected in the fact that in cases like the above, *exercise* is not a part of these verbs' meanings in the same way that their superordinate *move* is. The superordinate *move* serves to illustrate partially the meaning of *walk* and *run* in the formula *to walk/run/swim is to move in some manner.* By contrast, the statement *to walk/run/swim is to exercise in some manner* says nothing about walking, running, and swimming events, in part because it applies to such semantically heterogeneous verbs as *row, dive, dance,* and *bike.* However, the relation seems more useful when reversed: Verbs like *walk, run, swim, row, dive, dance* and *bike* serve well to describe the meaning of *exercise* in the formula *to exercise is to walk, run, swim, row, dive, dance in some manner.*

This is similar to the relation between *animal, dog,* and *pet.* To define *dog* as a kind of *pet* is odd and certainly not informative; but to define *pet* via its role "subordinates" such as *dog, cat, goldfish, hamster* is quite meaningful.

If one examined the hierarchical trees in WordNet top-down only, the function relation would indeed not be easily distinguishable from troponymy. It is only when one checks the relation of the troponyms to their superordinates that the difference becomes evident. Conflating the function relation and troponymy, as is done in WordNet, is therefore partially justified.

Another difference between strict troponymy and the function relation is that while troponyms share syntactic features with each other, verbs linked by the function relation to a common superordinate are not necessarily syntactically similar. *Row* can be a form of exercise, but unlike manners of motion involving no vehicle, like *run, walk, jog,* it may be transitive, as in the phrase *row a boat. Bike* and *boat* are denominals incorporating the vehicle names and are necessarily intransitive. Because so many different activities can be "exercise" (or simply defined as such), there is no reason that the verbs expressing these activities should have similar syntactic properties or select for the same noun arguments.

In the verb lexicon, the function relation turns out to be pervasive. As another example, consider the verb *punish.* Some ways to punish someone are *to spank, thrash, banish, imprison, fine.* Other than their relation to *punish,* these verbs are semantically heterogeneous and do not share all of their syntactic properties in common, either. Not surprisingly, there is no *punish* class of verbs in a syntax-semantics based classification of verbs like Levin (1993). *Spank, beat, flog* are genuine troponyms of *beat,* and they need not function to punish. In the case of *imprison,* the punishment meaning is perhaps more salient, but people are also imprisoned while being held under suspicion for a crime or to simply in order to be kept away from others, when the imprisonment does not constitute punishment for a transgression. *Fine,* on the other hand, is a genuine troponym of *punish:* It cannot be assigned to another superordinate that does not have the sense of *punish.* This example indicates that there are hierarchies of verbs related purely by function and that are not intersective with troponymy hierarchies. Unlike in the case of *run* and *walk,* the relation is symmetric: The statement *to fine (someone) is to punish (someone) in some manner* is fine, as is a definition of *punish* by means of its subordinate *fine.*

Consider next the verbs *wave, nod,* and *shrug.* They denote the motion of a particular body part, and the statements *to wave/nod/shrug is to move (one's hand/head/shoulders) in some manner* is acceptable. But their predominant sense is that related to their function sense as verbs of gesturing: Not only is it felicitous to define *gesture* in terms of verbs like *wave, shrug, nod,* but the statement *to wave/nod/shrug is to gesture in some manner* seems very natural; so the function relation here is virtually indistinguishable from troponymy.

While we can distinguish hierarchical structures based on genuine troponymy from those involving a function relation, we saw that the two relations are very similar; moreover, for some verbs, their meaning as function hyponyms is more salient than for others. Such shades in the strength of relatedness cannot be straightforwardly represented, and conflating troponymy and the function relation often seems to be a practical necessity in a semantic network like WordNet.

5. MANNER VS. RESULT

A number of verbs encode what we called a function, which might be context- or situation-dependent. These verbs say nothing about the manner in which the event is carried out. By contrast, other verbs denote the manner, but say nothing about the purpose of the event. Besides verbs expressing the function or manner in which an action is carried out, English has many verbs that encode the result of an action, but not the manner of achieving this result.

For example, *shut* denotes an event that results in a closed entity. But the manner of shutting varies widely, depending on the entity that is being shut. Shutting a book is a very different event from shutting a box or a window; shutting a sliding window requires an action distinct from shutting a casement window or a pivoting window or a skylight or a car window (to say nothing of closing a window on a computer screen). For verbs like *shut,* the appropriate reading of the verb must be forced (or "coerced" (Pustejovsky, 1995)) by the noun that denotes the entity that is acted upon. Other examples of "result" verbs are *open, break, destroy, clean,* and *melt.*

In contrast to result verbs, a manner (of motion) verb like *run* denotes a specific kind of traveling motion that does not vary depending on who (or what) does the running. Webster's definition of *run*—"go faster than walk: to go steadily by springing steps so that both feet leave the ground for an instant in each step"—applies equally to people, chicken, and gazelles.

Rappaport Hovav and Levin (1998) have called result verbs like *shut* accomplishments, and manner verbs like *run* activities. Besides the presence of different prominent meaning components, these two verb types distinguish themselves in their syntactic behavior. Activity verbs typically occur intransitively (as unergatives), whereas accomplishments must have a direct argument; many causative accomplishment verbs also have an inchoative intransitive (unaccusative) counterpart:

(5)   She ran/walked (to the store). (Activity, unergative)
(6)   *She shut.
(7)   She shut the door. (Accomplishment, causative)
(8)   The door shut. (Inchoative, unaccusative)

There seem to be no verbs that encode both the result and the manner of an event (Rappaport Hovav and Levin, 1998); the two have to be expressed by two different lexical items. But result and manner can be referred to as being part of the same event. Result verbs must express additional manner in a biclausal structure (as in (9)-(10)); manner verbs can take a resultative (as in (11)):

(9)   She shut the door (by) sliding it to the left.
(10)   She slid the door to the left, shutting it.
(11)   She slid the door shut.

Both manner and result verbs have subordinates and form hierarchical structures. It seems felicitous to say that *to swagger/ruffle/prance/strut/sashay/limp/hobble/shuffle is to walk in some manner,* and that *to snap/slam/bang is to close (something) in some manner.* Manner and result verbs thus behave alike with respect to the hierarchial structuring and the manner formula.

6. MANNER ELABORATION AND SYNTACTIC BEHAVIOR

Levin's (1993) analysis of a large part of the English verb lexicon tests the hypothesis that semantic similarity is reflected in shared syntactic behavior. Verbs are classified according to the syntactic alternations they do or do not participate in, and the resultant groups clearly show semantic coherence. Close inspection of the verb classes reveals that, in many cases, the members include both "basic" verbs, after which the class is named, and verbs that appear to be manner subordinates (troponyms) of the basic verbs. In fact, the syntactic behavior that Levin ascribes to the class sometimes characterizes only the basic verb, but is not shared by the manner-elaborated members.

For example, Levin identifies a class of "put" verbs, whose members include not only *put* itself, but also *arrange, immerse, install, lodge, mount, place, position, set, situate, sling, stash, stow.* These share a range of syntactic behaviors, such as the inability to participate in the Locative Alternation:

(12)   She put/arranged/installed the box on the desk.
(13)   *She put/arranged/installed the desk with the box.

By contrast, members of the related, but distinct "spray/load" class participate in this alternation:

(14)   She sprayed/spritzed/squirted paint on the door.
(15)   She sprayed/spritzed/squirted the door with paint.

Levin (1993) cites another syntactic property of her "put" class, the inability to occur without a prepositional phrase (PP):

(16)   *I put the box.

Levin's sentences illustrating the syntactic properties of the class generally contain the verbs after which the class is named; in this case, it is the verb *put.* But in fact, the class members do not behave uniformly with respect to the PP-requirement:

(17)   *I placed/set/situated/stashed/slung the boxes.
(18)   I arranged/installed/stowed the furniture.
(19)   The curator mounted the paintings.

Unlike *put,* and like the verbs in (18-19), *immerse* can also appear without a PP when the location is known in the discourse:

(20)   We arrived at the river and I immersed/*put my feet.

This syntactic difference suggests that *put, place, set, sling, stash, and situate* may belong to a more fine-grained subclass, and *arrange, immerse, install, mount,* and *stow* to another one.

We suggest that the difference exists because *put, place, set, situate,* in contrast to *arrange, install, stow, mount,* do not say anything about a particular use of space or resulting configuration. Arranging, installing, mounting, and stowing all imply that the moved entities end up in a preplanned order or spatial design; this does not seem to be implied by *put, place, set,* and *situate.* The corresponding passives clearly express this result:

(21)   The furniture was arranged/installed/stowed.
(22)   The paintings were mounted.
(23)   My feet were immersed.

Position patterns with *put, set,* and *situate,* except in cases like (25-26), where a particular, planned arrangement is implied:

(24)   *The clerk positioned the books.
(25)   The painter positioned the models.
(26)   The general positioned his men.

Sling, unlike all the other verbs, has a clear manner-of-putting component that distinguishes it from the other class members. *Stash* is likewise semantically distinct from the members of this group, as it denotes hiding or storing, rather than mere placing or arranging in some configuration.

The data suggest that the additional meaning component found in verbs like *arrange, immerse, install, mount,* and *stow* is related to the greater syntactic privileges of these verbs, as opposed to the superordinate verbs *put, place, set, situate,* which lack this semantic element.

A more fine-grained analysis of Levin's verb classes, which separates out "basic" verbs from manner elaborations, further supports the hypothesis of the close interrelation between syntax and semantics. Specific manner elaboration seems to be reflected in syntactic distinctions.

6.1 Slide Verbs

Another example comes from the small class of *slide* verbs, comprised of *bounce, float, move, roll, slide.* The grouping seems justified because these verbs behave alike with respect to three major syntactic alternations: causative, conative, and middle. But the so-called Dative alternation singles out one class member, *move,* which does not participate in this alternation:

(27a)   Mary bounced/floated/rolled the ball to John.
(27b)   Mary slid/moved the bail to John.
(27c)   *Mary moved John the ball.

Again, the verbs that encode a manner of moving an entity show greater syntactic privileges than the unelaborated verb.

6.2 Verbs of Sending and Carrying

In other cases, it is the less elaborated verbs that appear in a wider range of syntactic constructions.

Another verb class identified by Levin (1993) and labeled "Verbs of Sending and Carrying" includes verbs like *send, airmail, convey, deliver, shift, transfer.* One shared syntactic property attributed to the class members is the inability to participate in the causative-inchoative alternation:

(28a)   We sent/airmailed/conveyed/delivered/smuggled the book to John.
(28b)   *The book sent/airmailed/conveyed/delivered/smuggled to John.

However, of the 23 verbs that Levin lists, four stand out in being exceptions:

(29)   The responsibility shifted/transferred/passed/returned to John.

The verbs in (29) simply encode the (possibly abstract) movement of an entity from one possessor or location to another. By contrast, the verbs that do not have this intransitive alternation encode in addition the manner of transfer. As in the case of the verbs of putting, syntactic behavior separates semantically less elaborate verbs from the more elaborate members of a putative class. In the case of the verbs of sending and carrying, in contrast to the verbs of putting, it is the semantically less elaborate verbs that show the wider syntactic privileges.

We do not have an explanation as to why the core members of a semantic verb class often show different syntactic characteristics than the verbs that constitute semantic elaborations of these verbs. Moreover, it is puzzling that the more "basic" verbs have a more restricted syntax in some cases, but enjoy greater syntactic privileges in other cases. The unelaborated verbs, like *put,* occur with much higher frequency in the language than their troponyms, and they are acquired by children before the manner verbs are learned. They also take the widest range of noun arguments, whereas the troponyms, with their more specialized meanings, are more selective with respect to the nouns they choose.

7. SUMMARY AND CONCLUSION

A semantic network like WordNet represents the meaning of words by connecting them via a handful of conceptual-semantic relations. Perhaps the most significant is the hyponymy relation, which builds hierarchical structures. Troponymy, as this relation is called in the verb lexicon, apparently relates verbs in terms of a manner elaboration. It not only serves as the main organizer of verb concept, but can be shown to be a salient aspect of speakers' mental representation of verb meanings. We examined in some detail the semantics of troponymy and established that the formula expressing the manner relation fits many verb pairs that are in fact semantically distinct.

Endnotes

1. This work was supported in part by NSF grant 11S98-05732.

2. Explicit glosses and example sentences have been added more recently to facilitate the many natural language processing applications of WordNet.

References

Cruse, D. A. (1986). *Lexical Semantics.* Cambridge: Cambridge University Press.
Fellbaum, C. (1990). The English verb lexicon as a semantic net. *International Journal of Lexicography,* 3, 278-301.

Fellbaum, C. (Ed.). (1998). *WordNet: An Electronic Lexical Database.* Cambridge, MA: MIT Press.

Fellbaum, C., & Miller, G. A. (1991). Semantic networks of English. *Cognition,* 41, 197-229.

Fellbaum, C., & Chaffin, R. (1990). Some principles of the organization of the verb lexicon. *Proceedings of the 12th Annual Conference of the Cognitive Science Society,* 420-428. Hillsdale, NJ: Erlbaum.

Levin, B. (1993). *English Verb Classes and Alternations: A Preliminary Investigation.* Chicago: University of Chicago Press.

Miller, G. A. (Ed.). (1990). WordNet. Special Issue of *International Journal of Lexicography,* 3.

Miller, G. A. (1998). Nouns in WordNet. In C. Fellbaum (Ed.), *WordNet: An Electronic Lexical Database,* 23-46. Cambridge, MA: MIT Press.

Miller, K. J. (1998). Modifiers in WordNet. In C. Fellbaum (Ed.), *WordNet: An Electronic Lexical Database,* 47-67. Cambridge, MA: MIT Press.

Papafragou, A., Massey, C., & Gleitman, L. (2000). Shake, rattle, 'n' roll. Manuscript submitted for publication.

Pustejovsky, J. (1995). *The Generative Lexicon.* Cambridge, MA: MIT Press.

Rappaport Hovav, M., & Levin, B. (1998). Building verb meanings. In M. Butt & W. Geuder (Eds.), *The Projection of Arguments,* 97-134. Stanford, CA: CSLI.

Talmy, L. (1985). Lexicalization patterns: Semantic structure in lexical form. In T. L. Shopen (Ed.), *Language Typology and Syntactic Description,* 3:57-149. Cambridge: Cambridge University Press.

Chapter 3

Meronymic Relationships:
From Classical Mereology to Complex Part-Whole Relations

Simone Pribbenow

Center of Computing Technologies (TZI), Computer Science Department, University of Bremen, Bremen, Germany

Abstract:

Meronymic or partonomic relations are ontological relations that are considered as fundamental as the ubiquitous, taxonomic subsumption relationship. While the latter is well-established and thoroughly investigated, there is still much work to be done in the field of meronymic relations. The aim of this chapter is to provide an overview on current research in characterizing, formalizing, classifying, and processing meronymic or partonomic relations (also called part-whole relations in artificial intelligence and application domains). The first part of the chapter investigates the role of knowledge about parts in human cognition, for example, visual perception and conceptual knowledge. The second part describes the classical approach provided by formal mereology and its extensions, which use one single transitive part-of relation, thus focusing on the notion of "part" and neglecting the notion of (something being a) "whole". This limitation leads to classifications of different part-whole relations, one of which is presented in the last part of the chapter.

2. INTRODUCTION

Each entity in our world can be separated into parts: There are parts of objects, parts of events, and parts of abstract things like ideas, institutions, temporal or spatial entities. Although there are differences between parts of objects, parts of events, and parts of abstract entities, they also share a common core, as in the classification of part-whole relations or the notion of integrity of wholes.

There exists a broad range of different kinds of parts, in the domain of objects as well as in other domains. Components like the door of a house, the finish of a race, or the head of a company form the most "typical" kind of parts. Non-distinguishable lots of elements like grains in a sack of rice or seconds in an hour are parts that allow for an easier formalization. A formalization of part-whole relations must be able to deal with both of these kinds of pre-existing parts. Parts can also be constructed like the upper half of a car, the most thrilling part of a motion picture, or 100 grams out of a container of flour. Construction, selection, and measurement operators are needed to describe those kinds of parts.

A closer look at the ontological status of some parts reveals them as rather problematic entities: Is a keyhole really a "normal" part? It is a hole and therefore an entity that cannot exist independent of the rest of the lock. Nevertheless, this hole is the defining part of a lock. The ontological level also distinguishes between parts of individual entities and the part-hierarchy or partonomy of an object's concept, which is built up from the parts that normally constitute an individual of this concept. The latter can have essential and optional parts, thus representing the whole range of possible individual partonomies. Moreover, an entity, as well as the concept of this entity, can have different partonomies, depending on the conceptualization of the entity. A speech can be seen as a sequence of sounds on the phonological level or as a highly structured set of statements on the semantic level. The last ontological problem to mention here is the problem of (temporal) change. How can we describe the change in part-structure over time? Connected with (temporal) change is the problem of identity. Does an entity remain the same after all its parts have been changed?

In the following sections, we will address most of the topics mentioned in the introduction. In the second section, we will start with a review of the cognitive aspects of parts. In the third section, mereology—the "logic of parts"—is introduced, together with its limitations on formalizing common sense part-whole relations and approaches to overcome these problems. Based on these considerations, a classification of part-whole relations is described in the last section.

3. PARTS AND COGNITION

There is no doubt about the importance of parts for human cognition. Parts and their relations to each other play an important role in visual and auditorial perception. Reasoning about function, for example, about a device, is based on our functional knowledge about the parts of this device, and the order of inferences is guided by the spatial connections between the parts. Our vocabulary contains a variety of words for different kinds of parts or ways of partitioning a whole. The part-whole relation is important for natural language semantics, for example, for the interpretation of possessive constructions. The cognitive findings that we will present give further evidence how parts are embedded in (human) cognition.

In this section, we will concentrate on the parts of physical objects because this is a rather well-analyzed domain. A comparison between components of objects and those of events can be found in Rips & Estin (1998). Important findings about parts can be found mainly in the study of two areas: the human conceptual system and visual perception. As the notion of part is somewhat different in these two domains, we will present the theories of both domains and their interaction. Throughout this section, the same example, a standard desk lamp, will be used.

2.1 Parts in the Conceptual System

Following Tversky (1990), there are at least two ways of partitioning a complex concept: decomposition into subclasses and decomposition into parts. That leads to two different organizations of knowledge: The decomposition into subclasses leads to *taxonomies*, the one into parts to *partonomies*. While taxonomies and the taxonomic relation of class inclusion form a well established area of research, investigations concerning the meronymic relation of part-whole are relatively rare. This is surprising, because part knowledge seems to play an important role in cognition. Experiments by Litowitz & Novy (1984) give evidence that even 3-year old children possess meronymic knowledge and can communicate it using language. With increasing age, the children use more and more partonomic relations to describe objects. Younger children also have part-knowledge but do not use it as a primary method of description. Following Inhelder & Piaget (1964) and Markman (1989), children can process meronymic relations earlier than taxonomic relations.

Let us first focus on the similarities of taxonomic and partonomic relations as the two means for knowledge organization (see Rosch, 1977 and Mervis & Rosch, 1981 for taxonomies; Tversky & Hemenway, 1984 and Tversky, 1990 for partonomies and their relation to taxonomies). Both are asymmetric and transitive on the conceptual level (although there is no general transitivity for the meronymic relation, as we will see in the next section). They both form hierarchies consisting normally of three levels. Taxonomies are built up by the superordinate level (e.g., 'furniture'), the basic level (containing, e.g., the concepts 'table' and 'chair'), and the subordinate level (e.g., 'desk' and 'kitchen table' as specialized concepts of 'table'). Partonomies are built up from the whole object (e.g., 'desk'), its parts (e.g., 'top' and 'leg'), and sometimes subparts (e.g., 'drawers' as part of the part 'top'). Figure 1 shows an object with one possible conceptual partonomy.

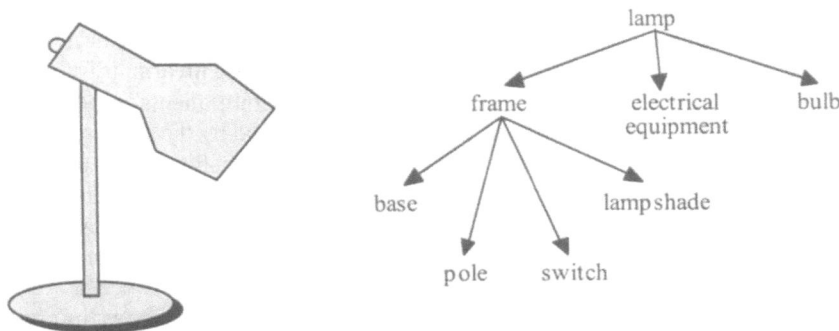

Figure 1. Partonomy of a desk lamp

In both taxonomies and partonomies, the middle level is the most important one for cognitive tasks. Parts, the middle level in partonomies, are used in visual perception (see Section 2.2) and for reasoning, especially in functional contexts. Basic level concepts, the middle level in taxonomies, are the first concepts to be learned by children; they are referred to by short words; and—most importantly—basic level concepts are associated with the greatest amount of knowledge.

Despite this structural correspondence there are differences in the processing of part-whole relations and class inclusion relations(Chaffin & Herrmann, 1989; Preuß & Cavegn, 1990). Class inclusion is based on similarity, and decisions on membership are made through feature comparison. Part-whole relations belong to our knowledge about the world, and decisions are made by looking up stored relations. Taxonomic relations are within concepts, while meronymic relations are between concepts. This distinction leads to differences in the possible inference processes. Class inclusion allows for inheritance of features from superordinate to subordinate concepts, while there is in general no inheritance from a whole to its parts. Instead, there is a kind of "upward" inheritance for special attributes like color, material, or function from parts to the whole (Tversky, 1990). For some attributes, for example, material, this upward inheritance is a simple addition of the attribute values of all parts; for attributes like function or form, additional rules for the combination of the parts' values are needed.

2.2 Parts in Perception and their Relations to Conceptual Parts

Up to now, there is no consensus about how objects are recognized. The identification of objects is a rather complex process, because people have to cope with a broad range of situations and variations. Objects must be identified at different locations, from different points of view, and in different lighting conditions. The shape of objects can vary in several ways: The shape of parts can be changed; the relations among parts can change for movable objects or for movable parts; optional parts may be included or not (Kosslyn, 1994).

Theories that try to explain object recognition using only template matching with (a set of) contour images cannot cope with the infinite variety of appearances described above. One possible explanation is to assume the use of visual invariants that are described in the object's concept. One, and probably the most important, of those invariants is the structure of components making up an object. The "Recognition-by-Components" model dates back to Biederman (1987). It claims that objects could be identified by the shades of their parts and the structural relations between them. Although the contour, that means the projection of the object, changes with motion of the object or a change of the view point, the general shade of parts and the attachment points among parts remain the same. The information on the component structure is stored in the objects' concepts in long term memory and contributes to the informativity of basic level concepts.

The question arising here is whether the perceptual parts are the same as the conceptual ones. Tversky and Hemenway (1984) claim that they are and that this fact is one reason for the importance of parts for cognition. The investigations of Schyns and Murphy (1994) show that there is no simple one-to-one relation as Tversky and Hemenway suggest. Perceptual parts are constructed from the two-dimensional projection of objects by using a principle proposed by Hoffman and Richards (1984). According to their *transversality principle*, two parts meet in a contour of concave discontinuity. To cope with continuous surfaces without sharp edges, this principle is generalized to the minima rule that divides a contour into its parts at the negative minima of each principal curvature along its associated family of lines of curvature. The arrows in figure 2 (left) indicate the segmentations predicted by this principle and the resulting perceptual partonomy (middle).

A comparison between the perceptual and conceptual partonomy reveals that they are not the same. First, all "hidden" parts of the conceptual partonomy are missing in the perceptual one. This observation is neither surprising nor problematic, because only the

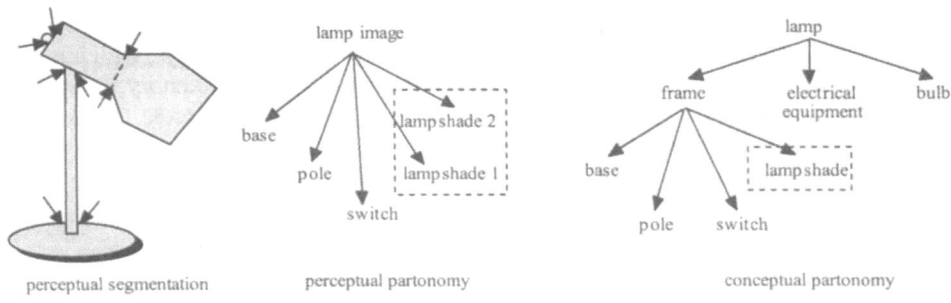

Figure 2. Result of the perceptual decomposition into parts

visual parts are used in recognition.   Second, the transversality principle divides the lampshade into two parts (see dotted square in the middle part), whereas people seem to disregard the segmentation and treat the lampshade as one undivided whole corresponding to the conceptual partonomy (dotted square in the right part). Third, different parts can be perceptually recognized as one, if they are aligned in the silhouette. A good example are the three parts of a finger that are no longer distinguishable when the finger is elongated.

Following Schyns and Murphy (1994), these phenomena prove that the transversality principle and minima rule do not fully determine the parts used in object recognition. These perceptual principles provide a robust basis for segmentation that indicate possible part boundaries. Further constraints are needed to determine those parts that are matched against conceptual parts. Based on experiments, the authors develop their *functionality principle* for part segmentation. According to the functionality principle, only those fragments are used in object recognition that enable the categorization. A fragment can be a perceptual part determined by the minima rule, a combination of two or more perceptual parts (as in the example of the lampshade), or even a part of a perceptual part (as in the example of the phalanxes of a finger). Fragments are normally contiguous. With respect to these findings, perceptual parts do not necessarily correspond one-to-one to conceptual parts, although there is a mapping between them guided by the functionality principle.

3. PARTS AND CLASSICAL EXTENSIONAL MEREOLOGY

Classical extensional mereology (CEM) is a familiar, and probably the most fundamental, approach towards a formalization of parts. It dates back to the work of Lesniewski (1916/1929) and the calculus of individuals of Leonard and Goodman (1940). The axiomatization given below is taken from Simons (1987) and Varzi (1996). The full formal system of CEM will be developed in three steps: basic mereology, extensional mereology, classical extensional mereology. Each step is followed by a discussion about the problems that arise if the axioms and their conclusions are applied to our common sense conception of part-whole relations. If applicable, the reader is referred to approaches toward overcoming these limitations.

3.1 Step 1: Basic Mereology

Basic mereology is based on a single relation part-of (x, y) or P (x, y), which is read as "x is part of y". Based on any calculus of first-order logic with identity, the part-of relation is described as reflexive (A1), asymmetric (A2), and transitive (A3). The proper-part-of relation PP (x, y) is defined as x being a part of, but not identical to y (D1). This asymmetric and transitive relation reflects better the linguistic and common sense notion of a "part".

(A1)	$P(x, x)$	reflexivity
(A2)	$P(x, y) \& P(y, x) \rightarrow x = y$	asymmetry
(A3)	$P(x, y) \& P(y, z) \rightarrow P(x, z)$	transitivity
(D1)	$PP(x, y) \rightarrow P(x, y) \& \neg P(y, x)$	definition of proper-part

For both the part-of and the proper-part-of relations, the axiom of transitivity is the target of criticism. Especially work in the domain of lexical semantics (compare, e.g., Lyons, 1977; Cruse, 1979) has shown that transitivity does not hold in general for part-whole relations. Two well-known counter examples are given in (1) and (2).

(1)   The house has a door.
      The door has a handle.
   →? The house has a handle.
(2)   The finger is part of the person.
      The person is part of the crowd.
   →? The finger is part of the crowd.

The problem of transitivity does not arise if only the basic space/time domain is taken into account. The spatial/temporal inclusion induced by the part-of relation is transitive, as is the mereological relation itself. In example (1), the problem is generated by the functional aspect that is not transitive. The semantics of "an object having a handle" implies that the handle is attached to the object and that the object can be moved using the handle. This interpretation is true for door/handle but not for house/handle. The functional or other special aspects that a part may provide for the whole are not captured by the part-of relation because the mereological relation ignores the role of the whole and any difference between various components of a whole. An analogue argumentation holds for example (2). A crowd is defined as an unstructured collection of persons, while a person is seen from a bodily point of view as a structured complex of body parts, for example, fingers. The transitive inference is not possible because several kind of part-whole relations are involved: the element-collection relation of person/crowd and the component-complex relation of finger/person. To solve the problem of transitivity, classifications of different part-whole relations have been developed that take into account important aspects of the relation of different parts with respect to the whole, thus extending the mereological part-of relation that only focuses on parthood (cf. Section 4). Although these classifications are a first step, they do not provide a complete solution to the problem of transitivity.

3.2 Step 2: Extensional Mereology

In order to extend the basic mereology given by axioms A1-A3 to an extensional mereology, the Principle of Supplementation, axiom A4, is added. One effect of this axiom is the property of extensionality (A4′). If x has at least one proper part and all proper parts of x are proper parts of y (and vice versa), then x and y are identical. The principle also ensures that if an entity has a proper part it has more than one (known as the weak Principle of Supplementation).

(A4) $\neg P(x, y) \rightarrow \exists z (P(z, x) \& \neg O(z, y))$ principle of supplementation

(A4′) $\exists z\, PP(z, x) \& \forall z (PP(z, x) \rightarrow PP(z, y)) \rightarrow x = y$ extensionality

(D2) $O(x, y) \rightarrow \exists z (P(z, x) \& P(z, y))$ definition of overlap

The property of extensionality caused by the supplementation principle ensures that entities are completely defined by their parts. It both levels out entities people tend to distinguish and diversifies entities people tend to conceptualize as one and the same. Therefore, extensionality does not hold for structured entities like objects, events, and most abstract entities like institutions. Applying extensionality, the words "no" and "on" are identical because they share the same proper parts, the letters "o" and "n". The structure, here the sequence of letters, is not taken into account. This problem can be solved by using several kinds of part-whole relations (cf. the problem of transitivity above) and representing the additional spatial, temporal, functional, or other structure of the parts with respect to a whole for complex part-whole relations. This solution is used in most knowledge representation approaches to part-whole relations, for example, Artale, Franconi, Guarino, & Pazzi (1996), Padgham & Lambrix (1994), and Sattler (1996).

The second problem with extensionality is that of differentiating one entity into different ones for each part that is replaced or lost. As a result, the object or event changes over time. One solution to this problem is the introduction of temporally or modally modified predicates (see Simons, 1987 for further readings).

3.3 Step 3: Classical (Extensional) Mereology

The last feature added to gain full CEM is the existence of a unique sum for arbitrary entities. Some authors extend the formalism further by introducing unique product and difference. The weak version introduces with the binary sum principle a unique sum (product, difference) for two entities, where the uniqueness is guaranteed by the property of extensionality implied by the principle of supplementation given above. The strong general form claims the existence of a sum for any arbitrary, non-empty set of entities. As the notion of a set or a class of elements is not available, the formalization of the general sum principle or fusion axiom (see AS5 below) is given via an axiom schema. Axiom schemata use predicate variables (in addition to the "normal" variables for domain objects), thus representing an infinite bundle of axioms of the same form, each with a different concrete predicate instantiation. In AS5, the predicate variable **F** stands for any property or condition that describes the set of entities to be summed up. Informally, the entity z in AS5 can be regarded as the sum of all entities denoted by the current instantiation of **F**.

Based on this principle, we can define an operator s for summing up arbitrary sets of entities. As for the sum of two entities, the general sum operator σ yields a unique result. In definition D3, this is indicated by the definite description operator ι, which is assumed to be a part of the logic vocabulary extending our first order logic.

(AS5) $\exists x\ F(x) \rightarrow \exists z \forall y\ (O\ (y, z) \rightarrow \exists x\ (F(x)\ \&\ O\ (y, x)))$ general sum principle

(D3) $\sigma x\ F(x) \rightarrow \iota z \forall y\ (O\ (y, z) \rightarrow \exists x\ (F(x)\ \&\ O\ (y, x)))$ definition of sum operator

The possibility of summing up arbitrary entities to form new objects is thought of as one of the most problematic features of mereology. For example, it allows summing up the smile of Mona Lisa, my red screwdriver, and the first three seconds of the new millennium to form a new entity as good as any other. The general sum principle increases dramatically the number of entities of the domain and is contrary to human cognition. People only accept the summation of entities if the resulting sum can be classified by their conceptual systems or serve a special purpose in a concrete situation. In the first case, the sum corresponds to a meaningful whole with its own name and concept, for example, the sum of the frame, the bulb, and the electrical equipment can be conceptualized as a lamp, while the sum of the bulb and the base does not correspond to a concept. In the second case, entities are grouped together, for example, to provide the target of a linguistic plural reference as in "two of the children of my sister".

One solution to the problem of meaningless sums is to forbid the construction of arbitrary sums. The other possibility is to take the concept of a whole seriously and introduce the notion of *integrity*. Integrity is a measure for the grade of wholeness that can indicate why the sum of frame, bulb, and electrical equipment forms a good whole (the lamp) whereas the sum of base and bulb does not. Eschenbach and Heydrich (1995) add the non-mereological primitives of precedence or region to CEM in order to define integrity for temporal or spatial entities, respectively. But their approaches are defined only for abstract entities and not for everyday objects or events. Varzi (1996) develops different combinations of mereology and topology and uses them to describe parthood and wholeness. He uses topological means to separate good sums from bad ones in the domain of spatial-temporal entities, stating that maximally strong self-connected objects form good wholes, whereas scattered ones or objects that are only externally connected as a box and its lid do not. Spatial-temporal connection forms a valuable indication of something being a whole (in the common sense notion) because many of our entities are maximally strongly self-connected, in the sense that each part is—internally—connected to one or more other parts of the same entity, but to no other entities. Nevertheless, Varzi's condition is too strong, because objects that are only externally connected—like all containers with separable lids or tops and scattered entities like archipelagos—are conceptualized as acceptable wholes by humans.

A broader but less formal approach to integrity is given by Simons (1987). He characterizes integrity as a gradated and aspectual concept. For example, Switzerland is seen as a confederate state, a rather loose whole, in the political aspect and as a continuous geographical entity, an integral one, in the spatial aspect. An integral whole is defined as some partition (Simons calls it division) into parts that—completely or to a satisfying extent—fulfills the condition that every member of that partition stands in a certain relation to every other member, and no member has this relation to any other thing outside the partition. This relation is called the characteristic relation R. Applying this definition, a car with its driver is a highly integrated whole with respect to a spatial-functional connection. All parts are spatially connected, at least via chains of other parts, and most of them are functionally connected, at least via comprehensive components, while there is nearly no connection to objects outside the partition that is spatial and functional at the same time, perhaps with the exception of the street. The motor of the car is a poorer whole because most parts have a connection to entities outside the partition.

3.4 Further Problems: Dependent Parts

Another problem mereology could not cope with is that of *dependent parts*. Dependent parts are those parts of an object that are non-material, like interiors or surfaces. They have linguistic references like "the interior of my drawer" or "the top of my table". Often they even provide essential functions for the whole object like a keyhole or a room. On the one hand, dependent parts, especially holes, behave like "normal" parts. They belong to a concept and provide a certain function and shape that is often typical for those kinds of objects. On the other hand, attributes like material, color, or weight do not apply to them and they allow for co-localization of other objects. A pencil could be stored in the interior of a drawer (which is an immaterial object) but not in the material parts of the desk. Most important, non-material entities like interiors or rooms exist only by virtue of their surrounding boundary objects. Once a wall is pulled down, two small rooms become one big one; similarly, there is no possibility of moving a keyhole and leaving the rest of the lock where it is.

Although there is no standard way to represent and reason about holes and other non-material objects, at least two different proposals can be found:

- Perception-based theories treat them as "negative parts" or "containers".
- The formal "hole theory" of Casati & Varzi (1997) treats them as special hole entities different from material objects.

In the basic theory of visual perception from Biederman (see Section 2.2), which describes all objects by their parts and their spatial relations, non-material parts like interiors are not mentioned. A cup is represented by a semiconcave cylinder as the body of the cup and a bended geon with a much smaller diameter attached to the side of the large geon as handle (fig. 3a).

In order to deal with these kinds of entities, Landau and Jackendoff (1993) introduce features to distinguish "negative" from normal positive geons. As shown in figure 3b, negative geons (grey part) are thought to be cut off from the positive geons (white parts)

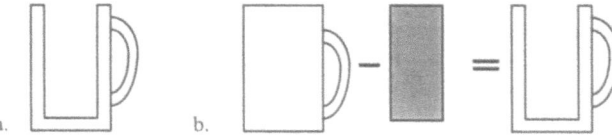

Figure 3. Several representation of a cup (cross-section), based on geons

of the object. Now, the body of a cup could be represented in two ways: like before, as a semiconcave positive cylinder (fig. 3a) or as a positive straight cylinder with a somewhat smaller and shorter negative geon cut out from the upper end of the bigger geon. From a cognitive point of view, the negative geon could be seen as the interior of the cup, thus allowing the mapping from perceptual to conceptual entities for non-material objects also. The disadvantage is that the positive basic geon for the body does not correspond to a conceptual part any longer. It is the *sum* of the positive and the negative body which relates to the conceptual body.

The second approach to non-material entities, the "hole theory" developed by Casati and Varzi (1997), covers topological holes as well as superficial hollows and internal cavities. In opposition to the perceptual theories, they treat holes like rooms or keyholes, not as parts in the mereological sense. Instead, they use a second primitive H (x, y), defined as "x is a hole in y", to represent holes. For example, a lock is described by the metal cylinder as material object o and the keyhole h as a hole associated with the material object o. The cognitive entity of a lock is represented as the "holed sum" s, consisting of the metal cylinder o and the keyhole h, $s = o + h$. In that construction, the keyhole h is part of the sum s but not of the object o. The hole theory classifies holes as immaterial spatial entities. Holes are represented by their specific primitive H (x,y) and are treated differently from "normal" substantial parts that way. Contrary to the approach of Landau and Jackendoff, which uses negative parts in modeling objects with holes, the hole theory does not produce entities without conceptual correspondence (see above). The disadvantage is that holes are treated differently from normal parts in *all* aspects. It is no longer possible to represent essential container objects like interiors or a keyhole as conceptual parts of an object. Neither an individual lock nor the concept of a lock includes a keyhole.

A common sense representation of objects with essential holes should try to catch the advantages of the two theories and avoid their disadvantages. An integrative approach is described in Engehausen, Pribbenow, & Töter (1996) and Pribbenow (1999), which uses an ontology and a hybrid propositional/diagrammatic representation to model objects. In that approach, holes are classified as immaterial entities with a special ontological status, thus following the hole theory. This special status, together with the use of diagrams, enables us to model holes as dependent objects whose spatial extent is induced by the surrounding material parts and which allow for co-localization. Following the theory of negative parts, holes should be parts, and objects should be allowed to contain non-material objects as parts. This assumption results in objects that are hybrid entities, which could consist of material *and* of immaterial parts. To avoid useless entities, holes are only constructed "on demand" on the conceptual level and not automatically by perceptual means, as the negative parts in the theory of Landau and Jackendoff. Conceptual considerations, normally functional ones, determine which holes are constucted, for example, the container provided by a cup, and which are left aside, for example, the hole between the body and the handle of the cup. The ontological status of holes is that of mental constructions and not of material objects; only attributes related to basic dimensions like shape and volume—but no other substantial attributes like weight or color—could be used to describe them.

4. CLASSIFYING PART-WHOLE RELATIONS

As Section 3 shows, classical extensional mereology does not capture our common sense notion of something being a part, because the mereological part-of relation neither captures the meaning of something being a whole nor provides possibilities for coping with the complex partonomies described in Section 2. Nevertheless, CEM provides a common core for further extensions (Simons, 1987; Eschenbach & Heydrich, 1995). In this section, we will present approaches to model more complex part-whole relations. Going beyond mereology, these approaches from artifical intelligence (AI) and cognitive science take care of the construction of wholes out of their parts and its effects for inferences on attribute values and transitivity. In this section, we discuss classifications of part-whole relations and possibilities to model the different part-whole relations.

The starting point for the development of classifications was the observation made by linguists that the part-whole relation is not in general transitive (see Section 3.1). Thus, for linguistic and conceptual tasks the simple 'part-of' relation of formal mereology is replaced by a family of relations. Winston, Chaffin, and Hermann (1987) and Iris, Litowitz, and Evens (1988) present well-analyzed classifications that were originally designed to deal with the problem of intransitivity between part-whole relations and to provide semantic primitives for describing part-whole relations. Their primarily linguistic motivation leads to classifications that are tailored to linguistic tasks, but not general enough for a general framework for parts and wholes (compare Gerstl & Pribbenow, 1995 for a detailed discussion of the—perhaps best known—approach by Winston, Chaffin, and Hermann). In the discipline of AI, other classifications have been developed for the use in knowledge-based systems, for example, Markowitz, Nutter, & Evens (1992), Sattler (1995) for an engineering application, Uschold (1996) for ecological information systems, or Bernauer (1996) for the medical domain. With the exception of the first one, these approaches focus on domain specific relations and do not claim to provide a general classification.

4.1 The Constructive Classification of Gerstl and Pribbenow

Gerstl and Pribbenow (1995) provide a general classification of part-whole relations that covers all ontological domains like physical objects, temporal and spatial entities, situations, and certain abstract entities, and is not limited to a specific domain. The basic idea of this classification is to give a *constructive* approach. Each part-whole relation represents a different way of partitioning a whole into parts. There are two alternative sources for parts:

- *Category A: The (a priori) structure of the whole*
  Most entities, like objects, events, or abstract entities like institutions, are structured into parts in a natural way, for example, objects into components, an event into its subevents. The natural part structure can change if another aspect is considered. From a material point of view, the book is structured into the cover and the single pages. With respect to content, it is structured into chapters. These permanent parts belong to the (conceptual) knowledge of the decomposed entity, i.e., the whole. Important for modeling these so-called "structure dependent" or

conceptual part-whole relations is the *(conceptual) domain knowledge* about *a priori* part structure of entities.

- *Category B: Partitions of the whole by construction*
  Parts could also be provided by segmentations of the whole based on internal attributes, like color for objects, or external schemes, like spatial frames. The resulting parts are temporary constructions and do not belong to domain knowledge. Important for modeling part-whole relations of that kind are the *construction processes* and the necessary knowledge about constructional means like attributes and schemes.

4.2 Characteristics of Conceptual Part-Whole Relations

Structure dependent or *conceptual part-whole relations* (category A) can be distinguished into three kinds of relations due to the complexity of the part structure of the whole: the *Component-Complex* relation, the *Element-Collection* relation, and the *Mass-Quantity/Set-Subset* relation.

The *Component-Complex* relation (CC) is used to represent a structure where each *a priori* part has a specific relation to the whole different from that of the other components. Examples of the CC relation from different domains are the decomposition of 'house' into its components 'roof' and 'main part', where each component plays a special role concerning (at least) spatial and functional aspects, or the decomposition of a complex activity like 'shopping' into its subevents 'moving to a shop', 'choosing an article', 'paying', etc., where each subevent has a different role with respect to time, resources, and intentions. The CC relation can be defined as a combination of the mereological proper-part-of relation (PP) described in Section 3 with the subrelations for every interesting aspect. For example, the CC relation for objects can be described by the combination of mereological, spatial, and functional relation. In general, every part p_i needs its own specific CC relation, denoted as CC_i, because p_i has a specific relation to the whole that is different—at least in one aspect—to the other part-whole relations. In addition, an integrity constraint for the whole is needed (cf. Section 3.3). A good choice for such a constraint is to claim that the object w serving as whole can be classified as belonging to a concept C. The integrity constraint remains the same for every CC relation because it represents a general condition for something being a good complex.

(Def. CC) $\forall p_i, w \ (CC_i \ (p_i, w) \rightarrow ((PP \ (p_i, w) \circ spatial_i \ (p_i, w) \circ functional_i \ (p_i, w))$
    $\& \ \exists C \ Concept \ (w) = C))$

The *Element-Collection* relation (EC) is used to describe a structure where each part has the same relation to the whole with respect to the relevant properties. This relation can be used to model a crowd consisting of—indistinguishable—persons or the different colors in the color menu of a computer program, under the condition that their ordering is not important. As there is no difference between the single parts, this relation can be modeled by one single part-whole relation. The EC relation is based on the mereological proper part-of relation (PP), together with additional restrictions on the sum operator. The concept of the whole defines which set of parts can form a sum of the appropriate kind. In order to sum up to a crowd of people, each single part must be a person and there must be enough of them. Both pieces of information, the classification of the parts and the additional sum condition, for example, on the number of parts needed, belong to (human) conceptual knowledge of possible collection entities.

(Def. EC) $\forall p,w$ ((EC (p, w) & Concept (w) = C & Has-elements (C, E)
     & Sum-Condition (w) = S) $\rightarrow$ (PP (p, w) & Concept (p) = E & S (w) = true)))

The *"Quantity-Mass-/Subset-Set"* relation (MQ/SsS) is used for structures that are restricted to the mereological proper-part-of structure, like a certain amount of wheat out of a sack of wheat (MQ) or two entities out of a set of arbitrary entities (SsS). The MQ relation is used for entities with a non-atomic basis like masses, while the SsS relation is used for sum objects built up from atomic parts. The whole has no (considered) structure or property in addition to the mereological part structure; the MQ/SsS relation can be reduced to the mereological proper-part-of relation.

(Def. MQ/SsS) $\forall p,w$ ((MQ (p,w) $\vee$ SsS (p, w)) $\rightarrow$ PP (p, w))

4.3 Constructed Parts

The second category of parts (B) consists of *temporarily constructed partitions*. If needed, constructed parts could be distinguished further with respect to the means used to single them out. We will use the notion of "Portions" for parts that are described by the values of their attributes. Examples for the partition of a whole into portions is to use the attribute "interestingness" to segment a novel into sections with respect to their degrees of entertainment value or to consider the attribute "color" to decompose an object into its red, blue, green, etc., parts. Construction processes for computing portions evaluate the structure of the value domain and the value distribution of the entity. The use of external schemes induces parts called *"segments"*. For example, the scheme of a one-dimensional directed line with its two ending points can be used to partition an entity into beginning, middle, and end phases or to select beginning, middle, or end points. Such a scheme can be applied to every entity that can be conceptualized or projected onto a line, for example, a river, a novel, or an abstract entity like a career. The construction processes for segments "implement" the different schemes and include procedures that determine whether a specific scheme can be applied to the entity under consideration.

Constructed parts could be identical with structure dependent parts (category A), but in general they are not. Temporary parts are constructed by processes that depend on the kind of partition (by internal attributes or external schemes, respectively) and the underlying representation. A detailed description of construction processes for hybrid logical/diagrammatic representations of objects can be found in Pribbenow (1999).

5. SUMMARY

This chapter analyzes and compares different views on parts and part knowledge. Section 2 presents experimental findings on the role of parts for the human conceptual system and visual perception. Conceptual knowledge about parts is organized in partonomies similar to taxonomies that are built up by class inclusion. These conceptual parts do not correspond directly to the perceptual parts used in object recognition. Perceptual parts are constructed by visual principles applied to the object's contour. Their relation to conceptual parts could be explained by the functional principle of categorization.

In Section 3, classical extensional mereology (CEM) is introduced as an established theory for formalizing parts. The axioms of this theory are discussed with respect to the common sense conception of part-whole relations reflected in the psychological findings. Limitations of CEM include the property of transitivity, which does not operate in some conceptual contexts; the property of extensionality, which does not hold for many structured entities and abstract entities; and the existence of arbitrary sums. CEM provides no means for some important features, e.g., to describe structural relations between parts like spatial or functional ones, to divide a "good" whole from a "bad" one, and to cope with "dependent" parts like holes or interiors.

Section 4 provides a classification of different part-whole relations that can solve some of the problems of CEM. It is a constructive approach that separates different conceptual part-whole relations from constructed parts and their relation to a whole. The conceptual relations are characterized based on CEM theory.

References

Artale, A., Franconi, E., Guarino, N., & Pazzi, L. (1996). Part-whole relations in object-centered systems: An overview. *Data and Knowledge Engineering, 20*, 347-384.

Bernauer, J. (1996). Analysis of part-whole relation and subsumption in the medical domain. *Data and Knowledge Engineering, 20*, 405-415.

Biederman, I. (1987). Recognition-by-components: A theory of human image understanding. *Psychological Review, 94*, 115-147.

Casati, R., & Varzi, A. (1997). Spatial entities. In O. Stock (Ed.), *Spatial and Temporal Reasoning*, 73-96. Dordrecht: Kluwer.

Cruse, D. A. (1979). On the transitivity of the part-whole relation. *Journal of Linguistics, 15*, 29-38.

Engehausen, A., Pribbenow, S., & Töter, U. (1996). Multiple part hierarchies. In F. Baader, H.-J. Bückert, A. Günter, & W. Nutt (Eds.), *Proceedings of the Workshop on Knowledge Representation and Configuration* (WRKP '96), 17-22. DFKI Document D-96-04.

Eschenbach, C., & Heydrich, W. (1995). Classical mereology and restricted domains. *International Journal of Human-Computer Studies, 43*, 723-740.

Gerstl, P., & Pribbenow, S. (1995). Midwinters, end games, and body parts: A classification of part-whole relations. *International Journal of Human-Computer Studies, 43*, 865-889.

Guarino, N., Pribbenow, S., & Vieu, L. (Eds.). (1996). Special issue on Part-whole relations. *Data and Knowledge Engineering Journal, 20*(3).

Hoffman, D. D., & Richards, W. A. (1984). Parts of recognition. *Cognition, 18*, 65-96.

Inhelder, B., & Piaget, J. (1964). *The Early Growth of Logic in the Child: Classification and Seriation*. London: Routledge & Kegan Paul.

Iris, M. A., Litowitz, B. E., & Evens, M. (1988). Problems of the part-whole relation. In M. Evens (Ed.), *Relational Models of the Lexicon*, 261-288. Cambridge, MA: Cambridge University Press.

Kosslyn, S. (1994). *Image and Brain*. Cambridge, MA: Harvard University Press.

Landau, B., & Jackendoff, R. (1993). "What" and "where" in spatial language and spatial cognition. *Behavioral and Brain Sciences*, 16, 217-238.

Leonard, H. S., & Goodman, N. (1940). The calculus of individuals and its uses. *Journal of Symbolic Logic*, 5, 45-55.

Lesniewski, S. (1929). Grundzüge eines neuen Systems der Grundlagen der Mathematik. *Fundamenta Mathematicae*, 14, 1-81. (First published in Polish in 1916.)

Litowitz, B. E., & Novy, F. A. (1984). Expression of the part-whole semantic relation by 3- to 12-year-old children. *Journal of Child Language*, 11, 159-178.

Lyons, J. (1977). *Semantics I*. Cambridge: Cambridge University Press.

Markman, E. M. (1989). *Categorization and Naming in Children*. Cambridge, MA: MIT Press.

Markowitz, J., Nutter, T., & Evens, M. (1992). Beyond IS-A and part-whole: More semantic network links. In F. Lehmann & E. Y. Rodin (Eds.), *Semantic Networks in Artificial Intelligence*, 377-390. Oxford and New York: Pergamon Press.

Mervis, C., & Rosch, E. (1981). Categorization of natural objects. *Annual Review of Psychology*, 32, 89-115.

Padgham, L., & Lambrix, P. (1994). A framework for part-of hierarchies in terminological logics. In J. Doyle, E. Sandewall, & P. Torasso (Eds.), *Principles of Knowledge Representation and Reasoning*: Proceedings of the Fourth International Conference on Knowledge Representation ('KR 94), 485-496. San Francisco: Morgan Kaufmann.

Preuß, M., & Cavegn, D. (1990). Semantische Relationen und Wissensstrukturen. Experimente zur Erkennung der Unter-Oberbegriffs- und Teil-Ganzes-Relation. *Zeitschrift für Psychologie*, 198, 309-333.

Pribbenow, S. (1999). Parts and wholes and their relations. In G. Rickheit & C. Habel (Eds.), *Mental Models in Discourse Processing and Reasoning*, 359-382. London: Elsevier Science.

Rips, L., & Estin, P. (1998). Components of objects and events. *Journal of Memory and Language*, 39, 309-330.

Rosch, E. (1977). Human categorization. In N. Warren (Ed.), *Advances in Cross-Cultural Psychology*, vol.1, 1-49. New York: Academic Press.

Sattler, U. (1995). A concept language for an engineering application with part-whole relations. In A. Borgida, M. Lenzerini, D. Nardi, & B. Nebel, (Eds.), *Proceedings of the 1995 International Workshop on Description Logics*, 119-123. Rome: Università degli Studi di Roma "La Sapienza", Dipartimento di Informatica e Sistemistica.

Sattler, U. (1996). A concept language extended with different kinds of transitive roles. In G. Görz, & S. Hölldobler, (Eds.), *KI-96: Advances in Artificial Intelligence*, 333-345. Berlin: Springer Verlag.

Schyns, P. G., & Murphy, G. L. (1994). The ontogeny of part representation in object concepts. In D. Medin (Ed.), *The Psychology of Learning and Motivation*, vol. 31, 305-349. San Diego: Academic Press.

Simons, P. (1987). *Parts. A Study in Ontology*. Oxford: Clarendon Press.

Tversky, B. (1990). Where partonomies and taxonomies meet. In S. L. Tsohatzidis (Ed.), *Meanings and Prototypes: Studies in Linguistic Categorization*, 334-344. New York: Routledge.

Tversky, B., & Hemenway, K. (1984). Objects, parts, and categories. *Journal of Experimental Psychology: General*, 113, 169-193.

Uschold, M. (1996). The use of the typed lambda calculus for guiding naive users in the representation and acquisition of part-whole knowledge. *Data and Knowledge Engineering*, 20, 385-404.

Varzi, A. (1996). Parts, wholes, and part-whole relations: The prospects of mereotopology. *Data & Knowledge Engineering*, 20, 259-286.

Winston, M., Chaffin, R., & Herrmann, D. (1987). A taxonomy of part-whole relations. *Cognitive Science*, 11, 417-444.

Chapter 4

The Many Facets of the Cause-Effect Relation

Christopher Khoo, Syin Chan, & Yun Niu
Centre for Advanced Information Systems, School of Computer Engineering, Nanyang Technological University, Singapore

Abstract:
This chapter presents a broad survey of the cause-effect relation, with particular emphasis on how the relation is expressed in text. Philosophers have been grappling with the concept of causation for centuries. Researchers in social psychology have found that the human mind has a very complex mechanism for identifying and attributing the cause for an event. Inferring cause-effect relations between events and statements has also been found to be an important part of reading and text comprehension, especially for narrative text. Though many of the cause-effect relations in text are implied and have to be inferred by the reader, there is also a wide variety of linguistic expressions for explicitly indicating cause and effect. In addition, it has been found that certain words have "causal valence"–they bias the reader to attribute cause in certain ways. Cause-effect relations can also be divided into several different types.

1. WHAT IS CAUSATION?

The cause-effect relation affects all aspects of our lives. It pervades our thinking and motivates our rational actions. Knowledge of cause and effect provides the basis for rational decision-making and problem-solving. It is important in all areas of science and technology. It can be argued that the ultimate goal of most research is to identify cause and effect.

Hume (1740/1965) called the association of ideas, which he said underlies the concept of causation, the *cement of the universe*. Hitchcock (1998) referred to causal knowledge as the great guide of human life. Keil (1989) suggested that cause-effect relations are essential and more useful than other sorts of relations in governing the structure of concepts and intuitive theories:

> The tremendous cognitive efficiency gained by using causal connections as a kind of glue has been repeatedly demonstrated in other areas such as text comprehension and story understanding; but it is equally if not more evident with respect to single concepts. Causal relations make it vastly easier to remember the features that make up a concept as well as to make inductions about new instances. (Keil, 1989, p. 280)

The concept of causation is complex and multifaceted, and it is surprisingly difficult to define. Two philosophers who contributed a great deal to our understanding of causation are David Hume and John Stuart Mill.

For Hume (1740/1965), causation comprises the following three conditions:

- *Contiguity* in time and place,
- *Priority* in time, and
- *Constant conjunction* between the cause and the effect.

When a person finds, from experience, that an event of the kind A is always followed by an event of the kind B, the person comes to conclude that event A causes event B. For Hume, causation is nothing more than the association in the mind of two ideas as a result of experiencing their regular conjunction.

Mill (1872/1973) argued that constant conjunction is not sufficient for inferring causation, unless the conjunction is also unconditional. Mill described four methods by which one can determine that A causes B:

- *The method of agreement*: If two or more instances of a phenomenon B have only one circumstance A in common, then A is the cause or effect of B.
- *The method of difference*: If we compare an instance X in which a phenomenon B occurs and an instance Y in which the phenomenon does not occur, and we find that the two instances have the same circumstances except that a circumstance A occurs in X but not in Y, then A is the cause or the effect of B.
- *The method of residues*: Subtract from any phenomenon the part that is known to be the effect of certain antecedents, then the remaining part of the phenomenon is the effect of the remaining antecedents.
- *The method of concomitant variations*: If a phenomenon B varies in a particular way whenever another phenomenon A varies in some particular way, then A is a cause or an effect of B.

Perhaps the most influential of these ideas is *the method of difference*. According to this method, we can conclude that event A causes event B if we find two instances which are similar in every respect except that in one instance event A is followed by event B, whereas in the other instance both A and B do not occur. Mackie (1980) argued that the layman uses this kind of reasoning to infer cause-effect relationships. In deciding whether a particular event A caused an event B, we engage in the counterfactual or contrary-to-fact reasoning that involves asking whether B would have occurred if A had not occurred. If B would not have occurred had A not occurred, we conclude that A caused B.

Mill's *method of difference* has been extended to distinguish between *necessary* and *sufficient* causes. An event of the kind A is a *sufficient* though not a *necessary* condition for an event of the kind B to occur if, when A occurs, B always follows, but when A does not occur, B sometimes occurs and sometimes not. On the other hand, if when A does not occur, B never occurs, but when A occurs, B sometimes occurs and sometimes not, then A is a *necessary* though not a *sufficient* condition for B to occur.

Mill (1872/1973) also pointed out that an effect is usually the result of a conjunction of several causes, even though in practice one of these causes is singled out as *the cause* and the rest are referred to as *conditions*. He said that "if we do not, when aiming at accuracy, enumerate all the conditions, it is only because some of them will in most cases be understood without being expressed, or because for the purpose in view they may without detriment be overlooked" (pp. 327-329). Mackie (1980) referred to the background conditions that are understood without being expressed as *the causal field*.

But is *cause* a necessary condition, a sufficient condition, or both? Mackie (1980) suggested that a cause is an *Insufficient* but *Necessary* part of an *Unnecessary* but *Sufficient* condition for an event. This is often referred to as the *INUS condition*. Jaspars, Hewstone, and Fincham (1983) and Jaspars (1983) found evidence that whether a cause is a necessary or sufficient, or necessary and sufficient, condition varies with the type of entity being considered for causal status. Cause is likely to be attributed to a person if the person is a sufficient condition. Necessity does not appear to be important when a person is a candidate for causal status. On the other hand, cause is likely to be attributed to the circumstances or situation if the situation is a necessary condition. Sufficiency is not so important for situational causes. Cause is ascribed to a stimulus when it is both a necessary and sufficient condition. So, "a personal cause is seen more as a sufficient condition, whereas situational causes are conceived primarily as necessary conditions" (Jaspars, Hewstone, & Fincham, 1983, pp. 16-17).

Mackie (1980) also pointed out that our concept of causation includes some presumption of a continuity from the cause to the effect, a causal mechanism by which the cause generates the effect. We conceive of the effect as being "fixed" by the cause.

Some philosophers have distinguished between general causation and singular causation (e.g., Ehring, 1997; Mellor, 1995). Whereas *general causation* refers to the causal tendency or cause-effect relation between two *types* of events over time, *singular or local causation* refers to the cause-effect relation between two *particular* events. It has been argued that local causation need not be an instance of a general causal law, and indeed a unique and unrepeated event can be causal. Furthermore, an event may be considered causal in a particular instance even if, over time, it is not found to be a necessary or sufficient condition for the effect event.

In view of the problem that a cause need not be necessary or sufficient for its effect, the concept of probabilistic causation has gained popularity. In this probabilistic view, an event of the kind A causes an event of the kind B if the occurrence of A makes the occurrence of B more likely (Hitchcock, 1998). This view recognizes the possibility of indeterministic causation—that there are instances where the causal mechanism is inherently probablistic, as in the field of quantum mechanics. Treatments of this idea can be found in Eells (1991) and Salmon (1984).

But how do ordinary people in their daily lives decide that there is a cause-effect relation between two events? Social psychologists working in the area of attribution theory have found that the human mind has a very complex and sophisticated mechanism for inferring cause and effect. Humans use empirical information (in the form of covariation information) in combination with world knowledge to construct a theory of the causal mechanism that produced the effect (Shultz, 1982; Alloy & Tabachnik, 1984; White, 1995). Social beliefs and probably also cultural factors influence the attribution of responsibility.

Researchers in social psychology have developed many models of how humans use various types of information to attribute cause. This includes the *inductive logic model* (Jaspars, Hewstone, & Fincham, 1983), the *abnormal conditions focus model* (Hilton & Slugoski, 1986), the *analysis of variance model* (Kelley, 1973), the *likelihood ratio model*

(Ajzen & Fishbein, 1975), the *linear combination model* (Downing, Sternberg, & Ross, 1985), the *probabilistic contrast model* (Cheng & Novick, 1992), and the *joint model* (Van Overwalle, 1997).

2. CAUSAL INFERENCE IN TEXT COMPREHENSION

The previous section examined the concept of causation. People perform causal inferencing automatically to make sense of events in the world and to guide their interaction with the world. Automatic causal inferencing has been found to be an important part of reading and text comprehension. Comprehension of text involves identifying relations between the various events, states, and ideas expressed in the text. This allows the reader to construct a coherent, connected representation of the text in the reader's mind. There is a substantial body of research indicating that identifying and inferring cause-effect relations between events is central to the comprehension of narrative text, i.e., stories (van den Broek, 1989; van den Broek, Rohleder, & Narvaez, 1996).

Keenan, Baillet, and Brown (1984) found that when giving subjects two sentences to read, where the first sentence specifies a cause for the event in the second sentence, reading times for second sentences steadily increased as causal relatedness between the pair of sentences decreased. This suggests that establishing causal relatedness between sentences is an important part of text comprehension. The difficulty in establishing causal relatedness between pairs of sentences appears to increase the time it takes to comprehend the second sentence. Zwaan, Magliano, and Graesser's (1995) study provided additional support for this conclusion.

Cause-effect relations between events and statements in a story affect how well the events and statements are recalled. The greater the number of causal connections an event or statement has to the rest of the text, the better it is recalled by the reader (Trabasso, Secco, & van den Broek, 1984; Trabasso & van den Broek, 1985; Fletcher & Bloom, 1988; van den Broek, Rohleder, & Narvaez, 1996). Furthermore, events and statements that lie along a causal chain connecting a text's opening to its final outcome are recalled better than those not on the chain (Black & Bower, 1980; Trabasso, Secco, & van den Broek, 1984; Trabasso & van den Broek, 1985; Fletcher & Bloom, 1988). Black and Bower (1980) and Trabasso and van den Broek (1985) found that proximity to the causal chain was a strong predictor of recall for story events.

Cause-effect relations also affect the perceived importance of story events. Events with more causal connections and that occur in the causal chain from the beginning to the end of the story are judged more important by readers and are also more likely to be used in a summary of the story (Omanson, 1982; Trabasso & Sperry, 1985; Trabasso & van den Broek, 1985). Van den Broek (1988) found that the judged importance of a goal statement increased strongly as its number of cause-effect relations increased, independent of its hierarchical level. The more causal connections the goals had, the more important they were judged to be.

Clearly, the causal structure of a text determines how statements in the text will be understood and remembered. Many of the cause-effect relations in text have to be inferred using information in the reader's short-term memory, long-term memory, and background knowledge (van den Broek, Rohleder, & Narvaez, 1996). Fletcher and Bloom (1988) and other researchers have suggested that the goal of narrative comprehension is to discover a sequence of causal links that connect a text's opening to its final outcome.

3. EXPLICIT EXPRESSIONS OF CAUSE-EFFECT IN TEXT

Though many of the cause-effect relations in text are implicit and have to be inferred by the reader, the English language actually possesses a wide range of linguistic expressions for explicitly indicating cause and effect.

Linguists have identified the following ways of explicitly expressing cause and effect:

- Using *causal links* to link two phrases, clauses, or sentences;
- Using *causative verbs;*
- Using *resultative* constructions;
- Using *conditionals*, i.e., "if . . . then . . ." constructions; and
- Using *causative adverbs, adjectives* and *prepositions.*

3.1 Causal Links

Altenberg (1984) classified causal links into four main types:

- The adverbial link, e.g., *so, hence, therefore;*
- The prepositional link, e.g., *because of, on account of;*
- Subordination, e.g., *because, as, since;*
- The clause-integrated link, e.g., *that's why, the result was.*

He presented a detailed typology of causal links and an extensive list of such linking words compiled from many sources.

An *adverbial link* is an adverbial which provides a cohesive link between two clauses, which may be two sentences. An *adverbial link* can be:

- *An anaphoric adverbial*, which has an anaphoric reference to the preceding clause, e.g., "There was a lot of snow on the ground. *For this reason* the car failed to brake in time."
- *A cataphoric adverbial*, which has a cataphoric reference to the following clause, e.g., "There was a lot of snow on the ground *with the result that* the car failed to brake in time."

Whereas an *adverbial link* links two clauses, a *prepositional link* connects cause and effect in the same clause. The prepositional phrase formed usually has an adverbial function, i.e., the preposition links a noun phrase to the clause, for example,

The car failed to brake in time because of the slippery road.

The words "because of" function as a phrasal preposition. Occasionally, the prepositional phrase modifies a noun phrase, e.g.,

The car crash, due to slippery road conditions, could have been avoided had the road been cleared of snow.

A link by *subordination* can be:

- *A subordinator*, e.g., *because, as, since, for, so.*
- *A structural link* marked by a non-finite *ing*-clause, e.g., *"Being* wet, the road was slippery."
- *A correlative comparative construction*, e.g., "There was *so* much snow on the road *that* the car couldn't brake in time."

Lastly, *clause-integrated links* form part of the *subject* or the *predicative complement* of a clause. When the linking words are the *subject* of a clause, Altenberg called it a *thematic link*, for example,

The car didn't brake in time. The reason was that there was a lot of snow on the road.

The linking words "the reason" function as the subject of the sentence. When the linking words form part of the *predicative complement* of a clause, it is called a *rhematic link*, for example,

The car accident was due to the slippery road.

The linking words "due to" form the first part of the phrase that functions as the complement of the verb "to be".

3.2 Causative Verbs

Causative verbs (also called *lexical causatives*) are verbs whose meanings include a causal element. Examples include the transitive form of *break* and *kill*. The transitive *break* can be paraphrased as *to cause to break*, and the transitive *kill* can be paraphrased as *to cause to die*.

Thompson (1987) divided causative verbs into three groups:

- Transitive causative verbs that also have an intransitive usage; for example, "x *breaks* y" is paraphrased as "x causes y to *break*".
- Causative verbs that do not have an intransitive usage; the transitive *kill* is paraphrased using the intransitive *die*, a different word: "X *kills* y" is paraphrased as "x causes y to *die*".
- Causative verbs that do not have any intransitive verb that can be used in the paraphrase; the past participle form of the causal verb is used instead: "X *butters* y" is paraphrased as "x causes y to be *buttered*".

How can causative verbs be distinguished from other transitive words that are not causative, e.g., *hit, kick, slap* and *bite*? It may be argued that all transitive action verbs are causative since an action verb such as *hit* can be paraphrased as *to cause to be hit*. Indeed, Lyons (1977, p. 490) said that there is a natural tendency to identify causality with agency and that causativity involves both causality and agency. Wojcik (1973, pp. 21-22) said that "all agentive verbs involve the semantic prime CAUSE at some level." Agents may be said to "cause" themselves to do things. The sentence "John intentionally broke the window" seems to entail "John caused himself to break the window". Wojcik considered all action verbs to be causative verbs in this sense.

However, most writers do not equate action verbs with causative verbs. Thompson (1987) argued that verbs like *hit*, *kick*, *slap*, and *bite* are not causative. She said that whereas causative verbs accept events and states of affairs as subjects, verbs like *hit* do not. Consider the following sentences:

(1a)   Oswald killed Kennedy by shooting at him.
(1b)   Oswald's accurate shooting killed Kennedy.
(2a)   John broke the vase by shooting at it.
(2b)   John's accurate shooting broke the vase.
(3a)   Tom hit the can by shooting at it.
(3b)   *Tom's accurate shooting hit the can.

Examples (1b) and (2b) show that the subject of the verbs *kill* and *break* can be an event. Examples (3a) and (3b) show that the subject of the verb *hit* can only be a person or an object. Thompson said that this is because for action verbs like *hit* the subject is, in a sense, not separable from the result of the action.

The following criteria have been proposed for distinguishing causative verbs from other transitive action verbs:

- Causative verbs accept events and states of affairs as subjects, whereas other action verbs accept only agents as subjects (Thompson, 1987).
- Causative verbs are transitive verbs that also have an intransitive usage where the subject of the verb has the patient role (Szeto, 1988).
- Causative verbs specify the result of the action, whereas other action verbs specify the action but not the result of the action (Szeto, 1988).

Khoo (1995) adopted the third criterion as a working definition of a causative verb in his analysis of verb entries in the *Longman Dictionary of Contemporary English* (1987) to develop a comprehensive list of causative verbs. He identified a total of 2082 causatives verbs, which he categorized into 47 types of results.

Levin (1993) provided a systematic and extensive classification of verbs based on their syntactic behavior. The verbs in each class share a semantic component, i.e., their meanings have something in common. Though the focus of her work was not on identifying causative verbs, nevertheless many of the verb classes do have a causal component in their meanings.

3.3 Resultative Constructions

A resultative construction is a sentence in which the object of a verb is followed by a phrase describing the state of the object as a result of the action denoted by the verb. The following examples are from Simpson (1983):

(4a)   I painted the car *yellow*.
(4b)   I painted the car *a pale shade of yellow*.
(4c)   I cooked the meat *to a cinder*.
(4d)   The boxer knocked John *out*.

In example (4a), the adjective *yellow* describes the color of the car as the result of the action of painting the car. In each of the four examples, the phrase in italics is the resultative phrase describing the result of the action denoted by the verb. The examples show that a resultative phrase can be an adjective, a noun phrase, a prepositional phrase, or a particle. Simpson (1983) said that the most common kind of resultative is where the resultative phrase is an adjective.

Simpson (1983) showed that some verbs that are normally intransitive can take an object if followed by a resultative phrase:

(5a)   I cried.
(5b)   * I cried myself.
(5c)   I cried myself *to sleep.*
(5a)   * I cried my eyes.
(5b)   I cried my eyes *blind.*

The objects in these three sentences have been called *fake objects* (Goldberg, 1991).

Some transitive verbs will take as object a noun phrase that they don't normally accept as object, if the noun phrase is followed by an appropriate resultative phrase as in the following examples:

(6a)   John drank the beer.
(6b)   *John drank himself.
(6c)   John drank himself *into the grave.*
(6d)   *John drank me.
(6e)   John drank me *under the table.*

In examples (6c) and (6e), the object of the verb *drink* is not the "patient" of the verb, i.e., it does not denote the thing that John drank.

An important question is whether all verbs will take an appropriate resultative phrase. In other words, for each verb, is there some resultative phrase (possibly one that nobody has thought of yet) that the verb will accept? And if it is true that some verbs can take a resultative phrase and other verbs can't, then how can these two classes of verbs be differentiated? Is there a systematic explanation for these two classes of verbs? These questions have not been satisfactorily answered.

3.4 Conditionals

"If . . . then . . ." conditionals assert that the occurrence of an event is contingent upon the occurrence of another event. Since the contingency of one event on another suggests a cause-effect relation between the two events, *if-then* constructions often indicate that the antecedent (i.e., the *if* part) causes the consequent (the *then* part).

It has been found that people sometimes interpret an *if-then* construction as a conditional and sometimes as a biconditional (i.e., "if and only if"), depending on the context. A conditional specifies that the antecedent is *sufficient* for the consequent to happen. A biconditional (i.e., "if and only if") specifies that the antecedent is both *necessary and sufficient* for the consequent to happen.

Whether *if-then* is interpreted as a conditional or biconditional depends on the background information available to the subject (Cummins, Lubart, Alksnis, & Rist, 1991; Hilton, Jaspars, & Clarke, 1990; Rumelhart, 1979). If the subject can think of other antecedents that can lead to the consequent, then the subject will interpret *if-then* as a conditional, otherwise the subject will interpret it as a biconditional. Here are two examples from Rumelhart (1979):

> *If you mow the lawn then I will give you $5.* (biconditional)
> *If you are a U.S. senator then you are over 35 years old.* (conditional)

We tend to interpret the first statement as expressing a biconditional relation ("if and only if"), whereas it seems more natural to interpret the second as asserting a simple conditional relation ("if, but not only if . . .").

Another factor that influences the interpretation of *if-then* constructions is the extremity or rarity of the consequent event. The more extreme or unusual the consequent, the more likely it is that the antecedent is judged by subjects to be necessary but not sufficient (Hilton, Jaspars, & Clarke, 1990). Kun and Weiner (1973) and Cunningham and Kelley (1975) found that extreme or unusual events such as passing a difficult exam or being aggressive to a social superior seem to require an explanation in terms of multiple necessary conditions (e.g., working hard and being clever, and being drunk and provoked). On the other hand, non-extreme and frequent events (e.g., passing an easy exam) could have been produced by many sufficient causes on their own (e.g., working hard or being clever).

If-then constructions do not always indicate a cause-effect relation. In the following example there is no cause-effect relation between the "if" and the "then" part of the sentence:

> *If you see a lightning, you will soon hear thunder.*

Though hearing thunder is contingent on seeing lightning, one does not cause the other. Seeing lightning and hearing thunder are caused by the same atmospheric event. An analysis of the different conditional constructions and their semantics can be found in Dancygier (1993).

3.5 Causative Adverbs, Adjectives and Prepositions

Some adverbs and adjectives have a causal element in their meanings (Cresswell, 1981). One example is the adverb *fatally*:

> *Brutus fatally wounded Caesar.*
> *Catherine fatally slipped.*

These can be paraphrased as:

> *In wounding Caesar, Brutus caused Caesar to die.*
> *Catherine slipped, and that caused her to die.*

The adjective *fatal* also has a causal meaning:

Caesar's wound was fatal.
Guinevere's fatal walk . . .

Other examples of causal adverbs cited by Cresswell are:

- Adverbs of perception, e.g., *audibly, visibly*;
- Adverbs that are marginally perceptual, e.g., *manifestly, patently, publicly, conspicuously*;
- Adverbs that involve the notion of a result whose properties are context dependent, e.g., *successfully, plausibly, conveniently, amusingly, pleasantly*;
- Adverbs that suggest tendencies, liabilities, disposition or potencies, e.g., *irrevocably, tenuously, precariously, rudely*;
- Adverbs that refer not to causes but to effects, e.g., *obediently, gratefully, consequently, painfully*; and
- Adverbs of means, e.g., *mechanically, magically*.

Causal adverbs and adjectives are not well studied, and a comprehensive list of such adverbs and adjectives has not been identifed.

Prepositions are also sometimes used in text to indicate cause-effect relations. Dirven (1995) classified causative prepositions into the following categories:

- Cause as proximity
  - As accompaniment: *with*, e.g., "tremble with fear", "irritated with her";
  - As target: *at*, e.g., "bridle at a remark", "angry at him";
  - As connection and path: *by*, e.g., "impressed by", "excited by".
- Cause as source
  - As separation from contact: *of*, e.g., "die of thirst", "die of boredom";
  - As separation from a point: *from*, e.g., "die from drugs", "shiver from the cold", "suffer from migraine";
  - As separation from a volume: *out of*, e.g., "got a kick out of", "drink too much out of nervousness".
- Cause as volume
  - As enveloping volume: *in*, e.g., "find pleasure in", "revel in";
  - As dispersion path: *about*, e.g., "delighted about", "apprehensive about";
  - As back-and-forth motion: *over*, e.g., "fell in disgrace over debts", "weep over".

Dirven (1997) also analyzed how prepositions are used to indicate emotions as cause and as effect.

4. IMPLICIT CAUSAL ATTRIBUTION OF VERBS

We have seen that cause-effect relations are expressed both implicitly and explicitly in text. Implicit cause-effect relations are inferred by the reader using information expressed in the text as well as background knowledge. However, implicit causality in text can also take the form of a subtle bias. Some linguistic expressions do not have a causal meaning that readers are consciously aware of, but nevertheless bias the reader towards assigning responsibility or blame to a participant referred to in the text.

In particular, some verbs have "causal valence"—they tend to assign causal status to their subject or object. Some verbs give the reader the impression that the cause of the event is the participant occupying the syntactic subject position of the sentence. Other verbs suggest that the cause of the event is the participant in the object position. This phenomenon has been referred to as the *implicit* or *inherent causality* property of verbs (Brown & Fish, 1983; Caramazza, Grober, Garvey, & Yates, 1977).

This implicit causal attribution can be made explicit by requiring the reader to determine whether an ambiguous anaphoric pronoun refers to the subject or object of the verb. Garvey and Caramazza (1974) had subjects complete sentences such as the following:

The mother punished her daughter because she ___

In completing the sentence, the subject automatically makes a choice as to whether "she" refers to mother or daughter. Garvey and Caramazza also asked subjects to supply responses to questions of the form

Why did the director criticize the actor?

In constructing a response to this question the subject decides whether the reason lies with the director or with the actor.

Garvey and Caramazza (1974) found that for the verbs *confess, join, sell, telephone, chase,* and *approach*, subjects tended to assign the pronoun and the reason for the event to the subject of the verb, for example,

The prisoner confessed to the guard because he <u>wanted to be released.</u>

For the verbs *kill, fear, criticize, blame, punish, scold, praise, congratulate,* and *admire*, subjects tended to assign the pronoun and the reason for the event to the object of the verb, for example,

The mother punished her daughter because she <u>broke an antique vase.</u>

Researchers have attempted to identify classes of verbs that tend to attribute causality in one direction or the other. Corrigan (1993) and Corrigan and Stevenson (1994) identified the following groups of verbs that have causal valence:

1. Experiential verbs
1.1. Experiencer-stimulus verbs
1.2. Stimulus-experiencer verbs
2. Action verbs
2.1. Actor verbs
2.2. Non-actor verbs

Experiential verbs describe someone having a particular psychological or mental experience. For some experiential verbs, such as *like* and *fear*, the subject of the verb takes the semantic role of *experiencer*, whereas the object of the verb has the *stimulus* role, for example:

John (experiencer) fears Bill (stimulus). (Cause is attributed to Bill)

These are termed *experiencer-stimulus verbs.* For other experiential verbs, such as *charm* and *frighten,* the subject of the verb takes the stimulus role, whereas the object of the verb has the experiencer role, for example:

John (stimulus) frightens Bill (experiencer). (Cause is attributed to John.)

These are termed *stimulus-experiencer verbs.* Several studies have found that most experiential verbs tend to attribute cause to the *stimulus* regardless of whether the stimulus occupies the subject or object position, whether the sentence is active or passive, and whether the participants are animate or inanimate (Au, 1986; Brown & Fish, 1983; Caramazza, Grober, Garvey, & Yates, 1977; Corrigan, 1988, 1992).

Action verbs describe events in which the participant in the subject position of the verb acts on the participant in the object position. The subject and object can be animate or inanimate. The subject of the verb takes the semantic role of *agent* or *actor*, and the object of the verb takes the role of *patient.* Brown and Fish (1983) presented data indicating that action verbs give greater causal weight to the subject of the verb. However, several other studies (Au, 1986; Caramazza, Grober, Garvey, & Yates, 1977; Garvey, Caramazza, & Yates, 1974/1975) found many action verbs that gave greater causal weight to the object.

Corrigan (1988, 1992) distinguished between two types of action verbs, which she called *actor verbs* and *non-actor verbs.* Actor verbs, e.g., *harm* and *help,* tend to attribute the cause to the actor (i.e., subject) of the verb, regardless of whether the subject and/or object are animate. She found that many of these verbs have derived adjectives referring to the subject (i.e., actor):

Actor verbs	Derived adjectives referring to the actor
defy	defiant
help	helpful
dominate	domineering

For the sentence "John defies Bill", the derived adjective *defiant* refers to the subject *John.*

Non-actor verbs, which either have no derived adjectives or have derived adjectives that refer to the object, tend to attribute cause to the object when the subject and object are both animate but have different social status. Non-actor verbs tend to attribute cause to the subject when the subject and object are both same-sexed humans named by proper nouns, when the subject and object are both inanimate, or when the subject is inanimate but the object is animate.

Garvey, Caramazza, and Yates (1974/1975) found that the implicit causal attribution of a verb can be modified or reversed by the following:

- Negating the verb; the sentence "the doctor *did not* blame the intern . . ." produced a smaller attribution bias towards the object than when the verb is not negated.
- Converting the sentence to passive voice; when sentences are passivized, there is a shift in the direction of causal attribution towards the surface subject of the verb.
- Changing the nouns occupying the subject and object position; Garvey et al. (1974/1975) suggested that the relative social status of the participants occupying the subject and object position influence the causal attribution. For the sentence,

The father praised his son . . .

causality tends to be imputed more often to the object than for the sentence

The son praised his father . . .

5. TYPES OF CAUSATION AND ROLES IN CAUSAL SITUATIONS

From the forgoing discussion, it is apparent that there are different types of causation and different types of causes. It may be important to distinguish between them in some situations and applications. The cause-effect relation is important in knowledge-based and expert systems, and researchers have attempted to model human causal knowledge and causal reasoning (e.g., Kamerbeek, 1993; Konolige, 1994; Ligeza & Parra, 1997; Nayak & Joskowicz, 1996). To model causal knowledge, we need to know which types of causation are important for the intended application and to distinguish between them in the knowledge base. Causal situations are complex, and we may need to know what aspects or roles in causal situations need to be represented in the knowledge base.

We have not come across an exhaustive typology of causation and causal situations, although different researchers have highlighted different types of causation and causes. Besides physical events and states, cause can refer to a wide range of phenomena including mental cause (human thinking and reasoning), psychological cause, teleological cause (purpose), and statistical laws.

Aristotle (1996) distinguished between four kinds of cause:

- *Material cause*: The material that an object is composed of can be seen as causing its existence.
- *Formal cause:* The form, pattern, or structure of an object can be said to cause its existence.
- *Efficient cause:* This is mechanical cause and refers to whatever causes an object to change, move, or come to rest.
- *Final cause:* This is teleological cause and refers to that for the sake of which the change occurs. For example, we walk for the sake of health, and so health can be seen as the final cause of the walking.

In a particular causal situation, all these causes can occur at the same time.

Teleological cause or *final cause* refers to a special type of causation where the cause, in a sense, occurs after the effect—a kind of backward causation. In human decision-making, a teleological cause is the intended effect of an action. For example, when a doctor prescribes a drug to treat a disease, it is because of the doctor's belief that the drug will cure the disease. An intended effect in the future is thus the cause of an event in the present. This is related to *mental causation*, since the cause is related to thought processes in the doctor's mind. Teleological cause is also seen as occurring in nature and can be understood in terms of natural selection. For example, the wing of a bird can be said to exist in order that the bird may fly. A recent analysis of teleology can be found in Koons (1998).

Terenziani and Torasso (1995) provided a taxonomy of what a cause or an effect can be. In their taxonomy, a cause or event can be one of the following:

- A *state* that persists and does not change over a period of time; or
- An *occurrence*, which can be subcategorized into
    - An *event*—an occurrence with a culmination or climax; or
    - A *process*—an occurrence that is homogenous and does not have a climax or an anticipated result (e.g., snowing).

Events and *processes* can be further categorized into those that have duration (occur over a period of time) or are momentary. Thus, an *event* can be:

- A *punctual occurrence* or *achievement*, i.e., the climax of an act (e.g., "John reaches the top"); or
- A *development* or *accomplishment*, i.e., an event that occurs over a period of time, ending with a climax (e.g., "John wrote the letter in an hour").

A *process* can be:

- A *punctual or momentary process* (e.g., "John coughed"); or
- A *durative process or activity* that endures over a period of time (e.g., "John walked").

The taxonomy thus makes use of the following dimensions:

- *Durativity:* Punctuality (occurs momentarily) versus temporal extension (occurs over a period of time);
- *Telicity:* Whether the situation has a climax or not, which distinguishes *events* (having a climax) from *processes* (no climax); and
- *Stativity:* States of affairs versus actions, which distinguishes *states* from *occurrences* (actions).

Cause and effect can also be categorized according to temporal considerations. In some applications it may be necessary, for example, to distinguish between causal situations where the effect persists after the end of the cause and situations where the effect ends when the cause ends. Terenziani and Torasso (1995) listed the following special types of cause-effect relations:

- *One-shot causation:* The presence of the cause is required only momentarily to allow the action to begin (Rieger & Grinberg, 1977).
- *Continuous causation:* The continued presence of the cause is required to sustain the effect (Rieger & Grinberg, 1977).
- *Mutually sustaining causation:* Each bit of cause causes a slightly later bit of the effect, and vice versa (Guha & Lenat, 1990, p. 241).
- *Culminated event causation:* The effect comes about only by achieving the culmination of the causal event (e.g., "run a mile in less than 4 minutes" causes "receive a prize").
- *Causal connection with a threshold:* There is a delay between the beginning of the cause and the beginning of the effect, and the effect is triggered only when some kind of threshold is reached.

Warren, Nicholas, and Trabasso (1979) identified four types of cause-effect relations in narrative texts:

- *Motivation*: The relation between a goal or intention of a person to an action taken to accomplish or further this goal/intention;
- *Psychological causation*: How an event brings about a person's emotion, goal, desire or some other internal state;
- *Physical causation*: The mechanical causation in the physical world between objects and/or people; and
- *Enablement*: The conditions that are necessary but not sufficient for an event to happen.

Dick (1997), in attempting to model the causal situation in a legal case, found it necessary to distinguish between the following types of cause and effect:

- *Distant* versus *proximate* (direct) cause;
- *Animate* versus *inanimate* agent;
- *Animate agent* versus *instrument* (the direct mechanical cause, or the tool used by the agent);
- *Volitive* versus *non-volitive* cause, i.e., intentional versus accidental cause (for animate agents only);
- *Active* versus *passive* cause;
- *Central* versus *peripheral* (or abstract) cause: peripheral cause includes purpose and reason for an action;
- *Explicit* versus *implicit* cause;
- *Aims* (intended but unrealized goals) versus *actual effect*.

6. CONCLUSION

We have presented a broad survey of the cause-effect relation from the perspectives of philosophy, psychology, and linguistics, with emphasis on cause-effect relations in text. The concept of causation is clearly multifaceted and complex. The definition of causation varies from situation to situation, and from one application to another. There are also many types of causation, and a causal situation has many aspects or roles that may be relevant to a particular application. Researchers in diverse research areas have investigated the use of cause-effect relations in their particular fields.

Researchers studying how people categorize things have found that cause-effect relations are important in determining the structure of natural categories and how people categorize things (e.g., Keil, 1989; Wattenmaker, Nakamura, & Medin, 1988). People make use of the cause-effect relation to infer attributes and the relative importance of attributes during the categorization process. According to Wattenmaker, Nakamura, and Medin (1988), categories derive their coherence not from overlapping characteristic properties but from the complex web of causal and theoretical relationships in which these properties participate.

Researchers have developed many kinds of formalisms and techniques to represent causal knowledge in expert systems and to simulate causal reasoning (e.g., Bree,

Hogeveen, Schakenraad, Schreinemakers, & Tepp, 1995; Castillo, Cobo, Gutierrez, Iglesias, & Sagastegui, 1994; Ligeza & Parra, 1997). Model-based expert systems that use a causal model of a domain for diagnosing and solving problems are more robust than systems that use heuristics (e.g., Artioli, Avanzolini, Martelli, & Ursino, 1996; Gonzalez & Chang, 1997). Researchers have also developed computerized knowledge-acquisition aids for eliciting causal knowledge from human domain experts (e.g., Charlet, Reynaud, & Krivine, 1996; Grundspenkis, 1998) as well as for mining causal knowledge from data (e.g., Glymour & Cooper, 1999).

Some researchers have attempted to develop computer programs to extract cause-effect information automatically from various kinds of text: narrative text (e.g., Bozsahin & Findler, 1992; Mooney, 1990), expository text found in text books and journal articles (e.g., Kaplan & Berry-Rogghe, 1991; Garcia, 1997), newspaper text (e.g., Khoo, Kornfilt, Myaeng, & Oddy, 1998), and short messages and explanations (e.g., Selfridge, Daniell, & Simmons, 1985; Joskowsicz, Ksiezyk, & Grishman, 1989).

Most of the studies have made use of knowledge-based inferencing to identify cause-effect relations in the text. The knowledge bases used were typically hand-coded, and it was difficult to scale them up for realistic applications or apply them in another domain. Some researchers (e.g., Garcia, 1997; Khoo, Kornfilt, Myaeng, & Oddy, 1998; Khoo, Chan, & Niu, 2000) have attempted to make use of linguistic clues to identify explicitly expressed cause-effect relations in text without knowledge-based inferencing.

With the increasing amount of text accessible on the Internet and the World Wide Web, we expect to see an increasing number of studies focusing on extracting causal knowledge from the Web for knowledge discovery, text summarization, and the development of knowledge-bases.

References

Ajzen, I., & Fishbein, M. (1975). A Bayesian analysis of attribution processes. *Psychological Bulletin, 82*, 261-277.

Alloy, L. B., & Tabachnik, N. (1984). Assessment of covariation by humans and animals: The joint influence of prior expectations and current situational information. *Psychological Review, 91*, 112-149.

Altenberg, B. (1984). Causal linking in spoken and written English. *Studia Linguistica, 38*, 20-69.

Aristotle. (1996). *Physics* (R. Waterfield, Trans.). Oxford: Oxford University Press.

Artioli, E., Avanzolini, G., Martelli, L., & Ursino, M. (1996). An expert system based on causal knowledge: Validation on post-cardiosurgical patients. *International Journal of Bio-Medical Computing, 41*, 19-37.

Au, T. K.-F. (1986). A verb is worth a thousand words: The cause and consequences of interpersonal events implicit in language. *Journal of Memory and Language, 25*, 104-122.

Black, J. B., & Bower, G. H. (1980). Story understanding as problem solving. *Poetics, 9*, 223-250.

Bozsahin, H. C., & Findler, N. V. (1992). Memory-based hypothesis formation: Heuristic learning of commonsense causal relations from text. *Cognitive Science, 16*, 431-454.

Bree, D. S., Hogeveen, H., Schakenraad, M. H. W., Schreinemakers, J. F., & Tepp, D. M. (1995). Conditional causal modeling. *Applied Artificial Intelligence, 9*, 181-212.

Brown, R., & Fish, D. (1983). The psychological causality implicit in language. *Cognition, 14*, 237-273.

Caramazza, A., Grober, E., Garvey, C., & Yates, J. (1977). Comprehension of anaphoric pronouns. *Journal of Verbal Learning and Verbal Behavior, 16*, 601-609.

Castillo, E., Cobo, A., Gutierrez, J. M., Iglesias, A., & Sagastegui, H. (1994). Causal network models in expert systems. *Microcomputers in Civil Engineering, 9*, 315-328.

Charlet, J., Reynaud, C., & Krivine, J.-P. (1996). Causal model-based knowledge acquisition tools: Discussion of experiments. *International Journal of Human-Computer Studies, 44*, 629-652.

Cheng, P. W., & Novick, L. R. (1992). Covariation in natural causal induction. *Psychological Review, 99*, 365-382.

Corrigan, R. (1988). Who dun it? The influence of actor-patient animacy and type of verb in the making of causal attributions. *Journal of Memory and Language, 27*, 447-465.

Corrigan, R. (1992). The relationship between causal attributions and judgements of the typicality of events described by sentences. *British Journal of Social Psychology, 31*, 351-368.

Corrigan, R. (1993). Causal attributions to the states and events encoded by different types of verbs. *British Journal of Social Psychology, 32*, 335-348.

Corrigan, R., & Stevenson, C. (1994). Children's causal attribution to states and events described by different classes of verbs. *Cognitive Development, 9*, 235-256.

Cresswell, M. J. (1981). Adverbs of causation. In H.-J. Eikmeyer & H. Rieser (Eds.), *Words, Worlds, and Contexts: New Approaches in Word Semantics*, 21-37. Berlin: Walter de Gruyter.

Cummins, D. D., Lubart, T., Alksnis, O., & Rist, R. (1991). Conditional reasoning and causation. *Memory & Cognition, 19*, 274-282.

Cunningham, J. D., & Kelley, H. H. (1975). Causal attributions for interpersonal events of varying magnitude. *Journal of Personality, 43*, 74-93.

Dancygier, B. (1993). Interpreting conditionals: Time, knowledge, and causation. *Journal of Pragmatics, 19*, 403-434.

Dick, J. P. (1997). Modeling cause and effect in legal text. In D. Lukose, H. Delaguch, M. Keeler, L. Searle, & J. Sowa (Eds.), *Conceptual Structures: Fulfilling Peirce's Dream: Fifth International Conference on Conceptual structures, ICCS'97*, 244-259.

Dirven, R. (1995). The construal of cause: The case of cause prepositions. In J. R. Taylor & R. E. MacLaury (Eds.), *Language and the Cognitive Construal of the World*, 95-118. Berlin: Mouton de Gruyter.

Dirven, R. (1997). Emotions as cause and the cause of emotions. In S. Niemeier & R. Dirven (Eds.), *The Language of Emotions: Conceptualization, Expression, and Theoretical Foundation*, 55-83. Amsterdam: John Benjamins.

Downing, C. J., Sternberg, R. J., & Ross, B. H. (1985). Multicausal inference: Evaluation of evidence in causally complex situations. *Journal of Experimental Psychology: General, 114*, 239-263.

Eells, E. (1991). *Probabilistic Causality*. Cambridge: Cambridge University Press.

Ehring, D. (1997). *Causation and Persistence: A Theory of Causation*. New York: Oxford University Press.

Fletcher, C. R., & Bloom, J. B. (1988). Causal reasoning in the comprehension of simple narrative texts. *Journal of Memory and Language*, 27, 235-244.

Garcia, D. (1997). COATIS, an NLP system to locate expressions of actions connected by causality links. In E. Plaza & R. Benjamins (Eds.), *Knowledge Acquisition, Modeling and Management: 10th European Workshop, EKAW '97 Proceedings*, 347-352.

Garvey, C., & Caramazza, A. (1974). Implicit causality in verbs. *Linguistic Inquiry*, 5, 459-464.

Garvey, C., Caramazza, A., & Yates, J. (1974/1975). Factors influencing assignment of pronoun antecedents. *Cognition*, 3, 227-243.

Glymour, C., & Cooper, G. F. (Eds.). (1999). *Computation, Causation, and Discovery*. Cambridge, MA: MIT Press.

Goldberg, A. E. (1991). A semantic account of resultatives. *Linguistic Analysis*, 21, 66-96.

Gonzalez, H., & Chang, P. S. (1997). Boiler tube failure root cause analysis advisory system. *Proceedings of the Twenty-Ninth Southeastern Symposium on System Theory*, 201-205.

Grundspenkis, J. (1998). Causal domain model driven knowledge acquisition for expert diagnosis system development. *Journal of Intelligent Manufacturing*, 9, 547-558

Guha, R. V., & Lenat, D. B. (1990). *Building Large Knowledge Based Systems*. Reading, MA: Addison-Wesley.

Hilton, D. J., & Slugoski, B. R. (1986). Knowledge-based causal attribution: The abnormal conditions focus model. *Psychological Review*, 93, 75-88.

Hilton, D. J., Jaspars, J. M. F. , & Clarke, D. D. (1990). Pragmatic conditional reasoning: Context and content effects on the interpretation of causal assertions. *Journal of Pragmatics*, 14, 791-812.

Hitchcock, C. R. (1998). Causal knowledge: That great guide of human life. *Communication & Cognition*, 31, 271-296.

Hume, D. (1965). *An Abstract of a Treatise of Human Nature*. Hamden, CT: Archon Books. (Original work published 1740).

Jaspars, J. (1983). The process of attribution in common-sense. In M. Hewstone (Ed.), *Attribution Theory: Social and Functional Extensions*, 28-44. Oxford: Basil Blackwell.

Jaspars, J., Hewstone, M., & Fincham, F. D. (1983). Attribution theory and research: The state of the art. In J. Jaspars, F. D. Fincham, & M. Hewstone (Eds.), *Attribution Theory and Research: Conceptual, Developmental and Social Dimensions*, 3-36. London: Academic Press.

Joskowsicz, L., Ksiezyk, T., & Grishman, R. (1989). Deep domain models for discourse analysis. In H. J. Antonisse, J. W. Benolt, & B. G. Silverman (Eds.), *The Annual AI Systems in Government Conference*, 195-200.

Kamerbeek, J. (1993). Generating simulators from causal process knowledge. *Proceedings of the 1993 European Simulation Symposium*, 579-584.

Kaplan, R. M., & Berry-Rogghe, G. (1991). Knowledge-based acquisition of causal relationships in text. *Knowledge Acquisition*, 3, 317-337.

Keenan, J. M., Baillet, S. D., & Brown, P. (1984). The effects of causal cohesion on comprehension and memory. *Journal of Verbal Learning and Verbal Behavior*, 23, 115-126.

Keil, F. C. (1989). *Concepts, Kinds, and Cognitive Development*. Cambridge, MA: MIT Press.

Kelley, H. H. (1973). The process of causal attribution. *American Psychologist,* 28, 107-128.

Khoo, C. S. G. (1995). Automatic identification of causal relations in text and their use for improving precision in information retrieval. Doctoral dissertation, Syracuse University.

Khoo, C., Kornfilt, J., Oddy, R., & Myaeng, S. H. (1998). Automatic extraction of cause-effect information from newspaper text without knowledge-based inferencing. *Literary & Linguistic Computing,* 13, 177-186.

Khoo, C., Chan, S., & Niu, Y. (2000). Extracting causal knowledge from a medical database using graphical patterns. *38th Annual Meeting of the Association for Computational Linguistics Proceedings of the Conference,* 336-343.

Konolige, K. (1994). Using default and causal reasoning in diagnosis. *Annals of Mathematics and Artificial Intelligence,* 11, 97-135.

Koons, R. C. (1998). Teleology as higher-order causation: A situation-theoretic account. *Minds and Machines,* 8, 559-585.

Kun, A., & Weiner, B. (1973). Necessary and sufficient causal schemata for success and failure. *Journal of Research in Personality,* 7, 197-207.

Levin, B. (1993). *English Verb Classes and Alternations: A Preliminary Investigation.* Chicago: University of Chicago Press.

Ligeza, A., & Parra, P. F. (1997). And/or/not causal graphs—A model for diagnostic reasoning. *Applied Mathematics and Computer Science,* 7, 185-203.

Longman dictionary of contemporary English (2nd ed.). (1987). Harlow, Essex: Longman.

Lyons, J. (1977). *Semantics.* Cambridge: Cambridge University Press.

Mackie, J. L. (1980). *The Cement of the Universe: A Study of Causation.* Oxford: Oxford University Press.

Mellor, D. H. (1995). *The Facts of Causation.* London: Routledge.

Mill, J. S. (1973). *A System of Logic, Ratiocinative and Inductive: Being a Connected View of the Principles of Evidence and the Methods of Scientific Investigation* (J. M. Robson, Ed.). Toronto: University of Toronto Press. (Original work published 1843).

Mooney, R. J. (1990). Learning plan schemata from observation: Explanation-based learning for plan recognition. *Cognitive Science,* 14, 483-509.

Nayak, P. P., & Joskowicz, L. (1996). Efficient compositional modeling for generating causal explanations. *Artificial Intelligence,* 83, 193-227.

Omanson, R. C. (1982). The relation between centrality and story category variation. *Journal of Verbal Learning and Verbal Behavior,* 21, 326-337.

Rieger, C., & Grinberg, M. (1977). The declarative representation and simulation of causality in physical mechanisms. *Proceedings of the Fifth International Joint Conference on Artificial Intelligence,* 250-256.

Rumelhart, D. E. (1979). Some problems with the notion of literal meanings. In A. Ortony (Ed.), *Metaphor and Thought,* 78-90. Cambridge: Cambridge University Press.

Salmon, W. (1984). *Scientific Explanation and the Causal Structure of the World.* Princeton: Princeton University Press.

Selfridge, M., Daniell, J., & Simmons, D. (1985). Learning causal models by understanding real-world natural language explanations. *The Second Conference on Artificial Intelligence Applications: The Engineering of Knowledge-Based Systems,* 378-383.

Shultz, T. R. (1982). Rules of causal attribution. *Monographs of the Society for Research in Child Development,* 47, 1-51.

Simpson, J. (1983). Resultatives. In L. Levin, M. Rappaport, & A. Zaenen (Eds.), *Papers in Lexical-Functional Grammar,* 143-157. Bloomington, Indiana: Indiana University Linguistics Club.

Szeto, Y.-K. (1988). The semantics of causative and agentive verbs. *Cahiers Linguistiques d'Ottawa,* 16, 1-51.

Terenziani, P., & Torasso, P. (1995). Time, action-types, and causation: An integrated analysis. *Computational Intelligence,* 11, 529-552.

Thompson, J. J. (1987). Verbs of action. *Synthese,* 72, 103-122.

Trabasso, T., & Sperry, L. L. (1985). Causal relatedness and importance of story events. *Journal of Memory and Language,* 24, 595-611.

Trabasso, T., & van den Broek, P. (1985). Causal thinking and the representation of narrative events. *Journal of Memory and Language,* 24, 612-630.

Trabasso, T., Secco, T., & van den Broek, P. (1984). Causal cohesion and story coherence. In H. Mandl, N. L. Stein, & T. Trabasso (Eds.), *Learning and Comprehension of Text,* 83-111. Hillsdale, NJ: Erlbaum.

Van den Broek, P. W. (1988). The effects of causal relations and hierarchical position on the importance of story statements. *Journal of Memory and Language,* 27, 1-22.

Van den Broek, P. (1989). The effects of causal structure on the comprehension of narratives: Implications for education. *Reading Psychology,* 10, 19-44.

Van den Broek, P., Rohleder, L., & Narvaez, D. (1996). Causal inferences in the comprehension of literary texts. In R. J. Kreuz & M. S. MacNealy (Eds.), *Empirical Approaches to Literature and Aesthetics,* 179-200. Norwood, NJ: Ablex Publishing.

Van Overwalle, F. (1997). A test of the joint model of causal attribution. *European Journal of Social Psychology,* 27, 221-236.

Warren, W. H., Nicholas, D. W., & Trabasso, T. (1979). Event chains and inferences in understanding narratives. In R. O. Freedle (Ed.), *Advances in Discourse Processes, Vol. 2: New Directions in Discourse Processing,* 23-52. Norwood, NJ: Ablex.

Wattenmaker, W. D., Nakamura, G. V., & Medin, D. L. (1988). Relationships between similarity-based and explanation-based categorization. In D. J. Hilton (Ed.), *Contemporary Science and Natural Explanation: Commonsense Conceptions of Causality,* 204-240. Washington Square, NY: New York University Press.

White, P.A. (1995). Use of prior beliefs in the assignment of causal roles: Causal powers versus regularity-based accounts. *Memory & Cognition,* 23, 243-254.

Wojcik, R. H. (1973). The expression of causation in English clauses. Doctoral dissertation, Ohio State University.

Zwaan, R. A., Magliano, J. P., & Graesser, A. C. (1995). Dimensions of situation model construction in narrative comprehension. *Journal of Experimental Psychology,* 21, 386-397.

PART II

Relationships in Knowledge Representation and Reasoning

Chapter 5

Internally-Structured Conceptual Models in Cognitive Semantics

Rebecca Green
College of Information Studies, University of Maryland, College Park, MD, USA

Abstract:
   The basic conceptual units of cognitive semantics—image schemata, basic level concepts, and frames—are internally structured, with meaningful relationships existing between components of those units. In metonymy, metaphor, and blended spaces, such internal conceptual structure is complemented by external referential structure, based on mappings between elements of underlying conceptual spaces.

1. STRUCTURE AND RELATIONSHIPS

Structure and relationships are inextricably interconnected. Wherever structure exists, relationships occur between the components of the structure. Similarly, wherever relationships exist, structure emerges. This inseparability of structure and relationality explains why a chapter on the internal structure of linguistic and conceptual complexes should take its place in a volume on the semantics of relationships.

The fundamental role played by structure in cognitive semantics helps distinguish it from an objectivist semantics. In this latter family of theories the following principles hold (Lakoff, 1987, p. 279):

- Concepts are either primitive or complex.
- Primitive concepts have no internal structure.
- Complex concepts are built up, through systematic principles of composition, of conceptual primitives.
- Only primitive concepts are directly meaningful.

In contrast, cognitive semantics recognizes directly meaningful concepts (basic-level and image-schematic concepts) with internal structure. "A central principle of cognitive semantics is that concepts do not occur as isolated, atomic units in the mind, but can only be comprehended . . . in a context of presupposed, background knowledge structures" (Clausner & Croft, 1999, p. 2). Consequently, the most primitive of cognitive semantics' concepts are not elemental, but structured; they incorporate relationships. Moreover, cognitive semantics proposes that such basic internally-structured, relational concepts are at the heart of human thought, perception, reasoning, imagination, and so forth.

From an objectivist semantics perspective, primitive, unstructured concepts, including concepts corresponding to basic nominal classes, may combine relationally to form complex, structured objects. Langacker, however, representing a cognitive semantics perspective, clarifies that nominal predications—which designate things—often have internal structure, as, for example, when nouns designate parts (e.g., *handle, elbow, top*) and wholes (e.g., *refrigerator, body, box*) or when they designate members (e.g., *bees, islands, trees*) or groups (e.g., *swarm, archipelago, forest*) (1987, pp. 194-197, 214-217; 1991, pp. 74-75). Nominal predications and relational predications may thus share the same basic internal structure, as is the case with *together*, represented in figure 1a, and *group*, represented in figure 1b. In both cases, E1, E2, and E3 stand for three individuals (together constituting a group) and R1, R2, and R3 stand for the interconnections between them. That which is bolded in the figures represents the salient part of the representation, that which is "profiled" by the respective word. In the case of *together*, a relational predication, the interconnections between the members are profiled; in the case of *group*, a nominal predication, the collection of members is profiled. (The dashed lines around E1, E2, and E3 indicate that their individual profiling is less pronounced that their collective profiling.) Such evidence supports the view that lexical items can routinely designate concepts with internal structure.

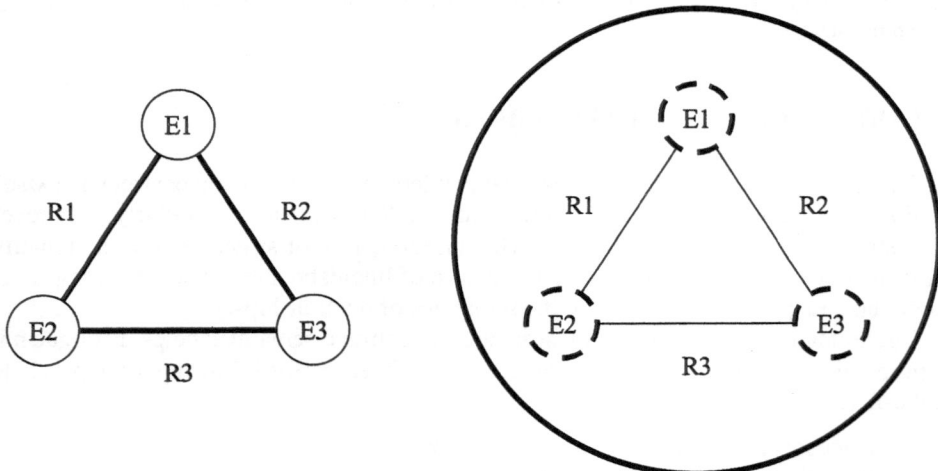

Figure 1a. Representation of *together*   Figure 1b. Representation of *group*

This chapter surveys the most prominent conceptual structures studied within cognitive semantics and demonstrates the ubiquity of internal structure at all conceptual levels. The discussion will begin by examining several basic cognitive semantic phenomena that are internally structured, namely, image schemata, basic-level concepts, and frame semantic structures. From here the discussion will move on to second-order conceptual structure phenomena that involve correspondences or mappings between the components of internally structured mental spaces. These phenomena include metonymy, conceptual integration (blended spaces), and metaphor and analogy. If the underlying spaces, which are generally organized by image schemata and/or frame semantic structures, were not internally structured, such mappings could not be established.

2. INTERNALLY-STRUCTURED CONCEPTUAL MODELS

In contrast to the tenets of an objectivist theory of semantics, which holds that primitive concepts have no primary structure, a basic premise of cognitive semantics is that the most primary concepts have a gestalt nature and therefore are internally structured. This will be seen to be the case with image schemata and basic level concepts, which are deemed by cognitive semantics to be directly meaningful concepts, inasmuch as they are experientially grounded, as well as with frames. Each of these phenomena are fundamentally relational in nature.

2.1 Image Schemata

Johnson (1987) characterizes image schemata as dynamic, continuous, and recurrent patterns that are based on our bodily experiences—for example, movement through space, manipulation of objects—and that structure our conceptions, perceptions, and actions. Not only is an image schema used for organizational purposes, but it also has internal organization. Indeed, it is its internal organization that allows an image schema to structure our interactions with the world: "A schema consists of a small number of parts and relations, by virtue of which it can structure indefinitely many perceptions, images, and events" (p. 29).

For example, one of the most commonly used image schemata is the CONTAINER schema. Its experiential bases include experiencing our own bodies as having emerged from a container (that is, our mother's womb), as well as experiencing our bodies as containers. The parts of the schema include a space interior to the container, a space exterior to the container, and a boundary between the interior and exterior. Another commonly used image schema is the JOURNEY schema, whose experiential basis is physical movement, and whose parts include a source point, a destination point, a path between them, and a traveler who traverses the path. The processual JOURNEY schema can be superimposed on the stative CONTAINER schema to give another processual schema, IN-OUT, in which a traveler moves along a path either from the interior of a container to the exterior of the container or from the exterior of a container to its interior, crossing the container boundary in both cases. Again we have the experiential basis of our bodies as containers, into which we take such 'travelers' as air, water, and food and from which we expel corresponding waste products.

Johnson exemplifies how common the IN-OUT image schema is, given the following activities that may occur at the beginning of a person's day:

> You wake *out* of a deep sleep and peer *out* from beneath the covers *into* your room. You gradually emerge *out* of your stupor, pull yourself *out* from under the covers, climb *into* your robe, stretch *out* your limbs, and walk *in* a daze *out* of the bedroom and *into* the bathroom. You look *in* the mirror and see your face staring *out* at you. You reach *into* the medicine cabinet, take *out* the toothpaste, squeeze *out* some toothpaste, put the toothbrush *into* your mouth, brush your teeth *in* a hurry, and rinse *out* your mouth. At breakfast you perform a host of further *in-out* moves—pouring *out* the coffee, setting *out* the dishes, putting the toast *in* the toaster, spreading *out* the jam on

the toast, and on and on. Once you are more awake you might even get lost *in* the newspaper, might enter *into* a conversation, which leads to your speaking *out* on some topic. (1987, pp. 30-31)

Clearly all *in*'s and *out*'s are not the same. Some are experienced directly as physical motion (e.g., walking *out* of the bedroom and *into* the bathroom, taking the toothpaste *out* of the cabinet, putting toast *in* the toaster), but many involve metaphorical projection of a concrete IN-OUT structure onto a more abstract domain (e.g., waking *out* of a deep sleep, brushing one's teeth *in* a hurry, entering *into* a conversation). Furthermore, Lindner (1981, 1982) has shown that there is not just one IN-OUT image schema, but a family of related image schemata. Focusing on constructions using *out*, she found three related image schemata. The first, OUT₁, is the image schema we have been describing, in which the traveler (Lindner's trajector, TR) moves from the interior of a confined area (the landmark, LM) to a position outside the boundary defining its area, as represented in figure 2a and as exemplified by *Pump out the air* and *Pick out the best theory*. The second, OUT₂, involves a reflexive trajector, initially coincident with the landmark boundary, that moves progressively outward, as represented in figure 2b and as exemplified by *Roll out the red carpet* and *Hand out the information*; the trajector moves to a position outside the *original* boundary; in essence, the boundary itself moves. The third, OUT₃, involves the trajector moving away from a single-dimension, point-like landmark rather than from a multi-dimensional one; it is represented by figure 2c and is exemplified by *The train started out for Chicago*. Despite their differences, the three image schemata share the same basic parts (interior, exterior, and boundary) and therefore much of the same relational structure.

Figure 2a. OUT₁   Figure 2b. OUT₂

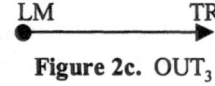

Figure 2c. OUT₃

The abstract nature of image schemata accounts for the widespread applicability of a reasonably small inventory of relational structures. Johnson's list of the most important image schemata numbers under thirty (1987, p. 126):

CONTAINER	BALANCE	COMPULSION
BLOCKAGE	COUNTERFORCE	RESTRAINT
REMOVAL	ENABLEMENT	ATTRACTION
MASS-COUNT	PATH	LINK
CENTER-PERIPHERY	CYCLE	NEAR-FAR

SCALE	PART-WHOLE	MERGING
SPLITTING	FULL-EMPTY	MATCHING
SUPERIMPOSITION	ITERATION	CONTACT
PROCESS	SURFACE	OBJECT
COLLECTION		

Many of these image schemata are discussed in Johnson, as well as in Lakoff (1987) and Krzeszowski (1993).

2.2 Basic Level Concepts

Many concepts are hierarchically related to other concepts in what is variously called a taxonomic, subsumptive, IS-A relationship. Typically, particular concrete entities can be said to belong to at least one class at the lowest (most specific) level of each appropriate hierarchy. If the hierarchical structure is indeed subsumptive, then those same entities will also belong to other classes at more general levels of the hierarchy. If an entity is appropriately classed at a number of hierarchical levels, does it make any difference from which level we derive its name? Or, as Roger Brown (1958) put the question, how shall a thing be called?

All other things being equal, we tend to call entities by names that designate classes toward the middle of the hierarchies they belong to. For instance, we are more likely to put on *shoes* and *socks* than *footwear* (a term designating a superordinate category) or *Oxfords* and *argyle socks* (terms designating subordinate categories). We sit on *chairs*, rather than on *furniture* or *high-backed chairs*. We take an *apple* to school for the teacher, not *fruit* or a *Stayman*.

This level of a hierarchy, called the basic level, is more than just the default level at which we name things. Lakoff (1987, p. 46), echoing Rosch, Mervis, Gray, Johnson, & Boyes-Braem (1976), summarizes a variety of ways in which this level is privileged. Specifically, basic level categories have the shortest names; their names are the earliest learned and understood by children and the earliest to enter the lexicon of a language; more of our knowledge about entities appears to be stored at the basic level than at any other level; and it is with basic level categories that people are most quickly able to identify category members. Why should basic level categories have this special status? Presumably this is because of three other characteristics of basic level categories: This is the "highest level at which category members have similarly perceived overall shapes, . . . at which a single mental image can reflect the entire category, [and] at which a person uses similar motor actions for interacting with category members." Ultimately, this boils down to the basic level being the highest level at which category members have the same parts configuration (Tversky & Hemenway, 1984; Tversky, 1986). In other words, the primacy of basic level concepts over superordinate categories is their shared internal structure, as defined by part-whole divisions. And it is the interaction between humans and the physical components of basic level categories that gives them their experiential basis and makes such concepts directly meaningful. Again, the most primary of concepts have internal structure, based especially in this case on part-whole relationships and on the relationships between parts and their functions.

2.3 Frames

The basic notion underlying frames is an "appeal . . . to structured ways of interpreting experience, . . . an alternative to the view that concepts or categories are formed through the process of matching sets of perceptual features" (Fillmore, 1976, p. 20). Elsewhere a semantic frame has been described as "any system of concepts related in such a way that to understand any one of them you have to understand the whole structure in which it fits" and "a system of categories structured in accordance with some motivating context" (Fillmore, 1982, pp. 111, 119). Thus, frame semantics is a theory of language understanding, situated within a world in which some types of experiences (e.g., events, conditions, relationships) recur frequently and are structured. Fillmore suggests that experiences are memorable "precisely because the experiencer has some cognitive schema or frame for interpreting it. This frame identifies the experience as a type and gives structure and coherence—in short, meaning—to the points and relationships, the objects and events, within the experience" (1976, p. 26).

The recurrence of prototypical experiences and the cultural or institutional meaning they have motivates the introduction of linguistic expressions to refer to the whole of these experiences, as well as to their salient parts or aspects. Such words or phrases are said to 'evoke' the frame, while the frame structures the meaning of those words and phrases (Fillmore, 1982, p. 117). While any given word will typically highlight (or profile) particular aspects of the frame, the whole of the frame structure is conveyed by each word or phrase that evokes it. This is essentially the point made by John Stuart Mill when he argues that the words *son* and *father* have the same connotation, at the same time that they clearly have different denotations:

> When we call one man a father, another a son, what we mean to affirm is a set of facts, which are exactly the same in both cases. To predicate of A that he is the father of B, and of B that he is the son of A, is to assert one and the same fact in different words. The two propositions are exactly equivalent: neither of them asserts more or asserts less than the other. The paternity of A and the filiety of B are not two facts, but two modes of expressing the same fact. That fact, when analysed, consists of a series of physical events or phenomena, in which both A and B are parties concerned, and from which they both derive names. What those names really connote, is this series of events: that is the meaning, and the whole meaning, which either of them is intended to convey. The series of events may be said to *constitute* the relation. (1843/1973, pp. 43-44)

To repeat a previous point: To understand any word that evokes a frame involves understanding the full frame structure, that is, the whole set of relationships that constitute the frame.

In the case of events, for which the richest frames exist, the internal structure of such prototypical scenarios typically includes some number of roles (that is, participants in the event, identified by the functions they play in the event), and may also include attributes of the event and subevents of the event. For example, Fillmore analyzed the JUDGING frame as consisting of:

a person who formed or expressed some sort of judgment on the worth or behavior of some situation or individual (. . . the Judge); a person concerning whose behavior or character it was relevant for the Judge to make a judgment (. . . the Defendant); and some situation concerning which it seemed relevant for the Judge to be making a Judgment (. . . the Situation). (1982, p. 116)

The Judge and the Defendant are examples of roles within the frame; the Situation is a subevent, requiring its own separate frame representation (depending on the semantic type of the situation being judged); attributes of the event include the degree of responsibility that the Judge ascribes to the Defendant for the Situation, as well as the nature of the Judge's judgment of the Situation—Was it positive or negative? In like manner, the COMMERCIAL TRANSACTION frame includes as its elements "a person interested in exchanging money for goods (the Buyer), a person interested in exchanging goods for money (the Seller), the goods which the Buyer could or did acquire (the Goods), and the money acquired (or sought) by the seller (the Money)" (Fillmore, 1982, p. 116). An inventory of representative frame structures for each of a dozen general domains is one of the outputs of the ongoing FrameNet (n.d.) Project, under the direction of C. J. Fillmore.

In a significant theoretical overview of frames, Barsalou proposes that they are "the fundamental representation of knowledge in human organization" (1992, p. 21); inasmuch as the various parts of a frame can also be described by frame structures, he further asserts that "human conceptual knowledge appears to be frames all the way down" (p. 40). Given the "combinability [of frames] into larger conceptual structures" (Fillmore, 1976, p. 30), it may be that human cognition involves frames all the way up as well. Barsalou's overall thesis is that significant psychological evidence points to the inadequacy of "flat", feature list representations of human cognition; frame representations not only avoid the problems encountered by feature list representations, but also support a wide array of cognitive processing tasks. Specifically, Barsalou argues that human knowledge evidences extensive use of attribute-value sets and relationships that are absent from feature list representations, but very much present in the basic components of frames, which include attribute-value sets, structural invariants, and constraints.

The first of these components—attribute-value sets—is the set of associations of values with specific slots of an instantiated frame. Barsalou notes that some of these attributes/slots can be said to constitute the frame's core, in that they frequently have specific values and may indeed be conceptually necessary: How, for example, can one have a COMMERCIAL TRANSACTION event without considering a buyer, a seller, merchandise, and payment? Other attributes/slots may occur in the representation of some frame instantiations, but not all.

The second of Barsalou's basic frame components—structural invariants—reflects the relatively invariant correlational and especially conceptual relationships between a frame's attributes/slots. He gives as examples the spatial relationship between the seat and back attributes of the CHAIR frame, the temporal relationship between the eating and paying attributes of the DINING OUT frame, the causal relationship between the fertilization and birth attributes of the REPRODUCTION frame, and the intentional relationship between the motive and attack attributes of the MURDER frame.

The third of Barsalou's basic frame components—constraints—also reflects relationships. In contrast to the normative relationships reflected by structural invariants, constraints involve "systematic variability in attribute values" in that "values constrain each other" (p. 37). For example, in the TRANSPORTATION frame, there is a negative attribute constraint between the values of the speed and duration attributes: The greater the speed of travel, the shorter its duration. In a similar vein, Fillmore andAtkins (1992) propose that various senses of verbs evoking the RISK frame can be distinguished according to the relationships they imply between attributes of the frame. For example, the meaning of *He risked death* involves a relation between the actor and harm, where *risk* can be paraphrased "to act in such a way as to create a situation of (danger for oneself)" (p. 99).

Barsalou's claim that human cognition is frames "all the way down" raises the question of primitives. As he poses the question, "Are there terminal components out of which all frames are constructed?" (p. 41). His answer notes that all of the basic components of a frame are susceptible to description involving internal structure:

> Although an attribute, relation, or constraint may start out as a holistic, unanalyzed primitive [that is, in our perception of it], aspects of its variability may subsequently be noted, represented with attribute-value sets, and integrated by structural invariants between attributes, and constraints between values. What was once a simple, unitary primitive becomes analyzed and elaborated, such that it becomes a complex concept . . . . Rather than being the elementary building blocks of knowledge, primitives may instead be larger wholes, the analysis of which produces an indefinitely large set of complex building blocks. (pp. 41-42)

Indeed, we should explicitly note that image schemata may be among the set of complex building blocks within a frame. For example, the prototypical COMMERCIAL TRANSACTION frame includes at least four instantiations of the JOURNEY image schema, two involving literal motion and two involving metaphorical motion. On the one hand, the payment makes a physical journey from the buyer's possession to the seller's possession, accompanied by a metaphorical passage of ownership of the money from buyer to seller; on the other hand, the merchandise makes a physical journey from the seller's possession to the buyer's possession, accompanied by a metaphorical passage of ownership of the merchandise from seller to buyer.

How, then, are frames and image schemata to be distinguished from each other? In terms of internal structure, there is no worthwhile distinction to be drawn. Their difference comes rather in those aspects of our experience they are used to reflect: We speak of "image schemata" when we are talking about recurrent patterns, for which we have an experiential basis; we tend instead to speak of "frames" when we are talking about stereotypical situations which have a psychological, social, or cultural basis. Clausner and Croft (1999) argue that image schemata are subsets of domains/frames and cannot be defined by a set of necessary and sufficient conditions. While image schemata are typically more abstract and schematic than frames and less complex, and while frames are more typically used to express propositional statements, together they share the characteristics of being fundamental and basic cognitive building blocks, all the while admitting of hypothetically unlimited internal elaboration and structure.

3. MAPPINGS ACROSS INTERNALLY-STRUCTURED CONCEPTUAL MODELS

Fauconnier (1997) suggests "that mappings between domains are at the heart of the unique human cognitive faculty of producing, transferring, and processing meaning" (p. 1). In contrast to the conceptual structure that is internal to the kinds of conceptual models we have just looked at, mappings rely on connectors that link such models externally and produce referential structure (Fauconnier, 1994, pp. x-xi; 1997, p. 39). All mappings reflect relationships.

Several classes of mappings may be distinguished. *Projection* mappings project part of the structure of one domain onto another, as occurs in metaphor. *Pragmatic function* mappings link between object categories that are related to each other in the real world, for instance, connecting authors to their writings or hospital patients to their illnesses. Metonymy is a type of pragmatic function mapping. *Schema* mappings connect the roles of general schemata or frames to their counterparts in mental space representations of their instantiations (Fauconnier, 1997, pp. 9-11). Complex schema mappings also occur in blended spaces.

We have already noted that image schemata and frames have the potential for complexity, not just because they have internal structure, but because some image-schemata and frames are structured recursively. On the one hand, an image schema may incorporate another image schema. For example, several FORCE schemata—COMPULSION, BLOCKAGE, COUNTERFORCE, RESTRAINT REMOVAL—include within them a PATH image schema; similarly, FULL-EMPTY is built on top of the CONTAINER image schema. On the other hand, a frame may incorporate both image schemata and other frames. For example, the COMMERCIAL TRANSACTION frame includes the JOURNEY image schema several times over; the Situation slot within the JUDGING frame could reference a number of different frames. However, despite the potential complexity engendered by these possibilities, we will still refer to image schemata and frames as first-order conceptual structures, inasmuch as they represent the least complex structures that are cognitively valid.

As we turn to the projection mappings of analogy and metaphor, and subsequently to schema mappings involving blends across mental spaces, we find ourselves dealing with necessarily higher-order conceptual structures, since these involve sets of correspondences across other conceptual structures, typically involving image schemata and frames. Pragmatic function mappings, however, are first-order conceptual structures, so we will take them up briefly first.

3.1 Metonymy

Metonymic mappings between categories of objects related to each other in the real world can typically also be conceived as mappings between wholes and parts within a frame or other cognitive model (since these capture typical real-world scenarios). However, metonymic mappings are not simply the kind of frame-internal relationships referred to in Section 2.3. They are specifically those mappings "in which one conceptual entity, the vehicle, provides mental access to another conceptual entity, the target, within the same domain" (Kövecses & Radden, 1998, p. 39); a "stands for" relation holds between the vehicle and the target (Lakoff, 1987, p. 78).

Common metonymies (with examples, taken from Lakoff & Johnson, 1980, pp. 36-39) include:

PART for WHOLE
   I've got a new *set of wheels.*
PRODUCER for PRODUCT
   He's got a *Picasso* in his den.
OBJECT USED for USER
   We need a better *glove* at third base.
CONTROLLER for CONTROLLED
   Napoleon lost at Waterloo.
INSTITUTION for PEOPLE RESPONSIBLE
   Exxon has raised its prices again.
PLACE for INSTITUTION
   Paris is introducing longer skirts this season.
PLACE for EVENT
   Watergate changed our politics.

While these examples commonly involve multiple image schemata or frames, the metonymies are each specific to a single image schema or frame. For example, the PRODUCER for PRODUCT metonymy occurs in the context of a PRODUCTION frame, whose slots include the PRODUCER, an agent who uses various MATERIALs and engages in various PROCESSes to create a resulting PRODUCT. In this metonymy, the value designating the PRODUCER slot can stand in for a value designating the PRODUCT slot. Metonymies are thus based on image-schematic or frame internal structure.

3.2 Metaphor

At its core, cognitive semantics envisions metaphor as "a cross-domain mapping in the conceptual system," realized through systematically related linguistic expressions (Lakoff, 1993, p. 203). The mapping is asymmetric, typically from a more concrete, source domain to a more abstract, target domain. The metaphorical mapping allows the target domain, with which a person generally has less experiential familiarity, to be understood in terms of the source domain, with which the person generally has greater experiential familiarity. The mapping includes both ontological correspondences, in which entities in the source domain are mapped onto entities in the target domain, and inferential correspondences, in which reasoning patterns from the source domain are mapped onto the target domain (pp. 206-208).

For example, the THEORIES ARE CONSTRUCTED OBJECTS metaphor (as it is identified in Lakoff, 1994) includes the following correspondences, as summarized by Grady (1997, p. 269):

- Major premises, including facts and assumptions, are the *foundation* of the theory.
- The major claims and arguments of a theory, along with their organization, make up its *framework.*

- Facts are solid materials or supporting elements—as in the sentence *Your facts are not solid enough to support your hypothesis.*
- Arguments are intermediary elements: They are *supported* by facts, they in turn *support* conclusions and claims.
- Claims are the uppermost elements of structure—they are *supported* by facts and arguments.
- The general logical structure is the *design.*
- A theoretician is an *architect.*
- The convincingness of the theory—its resistance to counter argument and disproof—is its *strength.* A theory may also be *flimsy*, and so forth.
- The continued existence of the theory as an accepted set of relevant claims and arguments is the duration of a physical structure; failure is understood as *collapse.*

Lakoff notes that "each mapping defines an open-ended class of potential correspondences" (1993, p. 210). The mappings, however, are partial: Some number of the potential correspondences are conventionally realized, i.e., are members of the established set of correspondences (*He is trying to buttress his argument with a lot of irrelevant facts, but it is still so shaky that it will easily fall apart under criticism*). Some number of the potential correspondences are subject to realization in novel contexts without becoming conventionalized (*Your theory is constructed out of cheap stucco*). And some number of the potential correspondences are never realized (*?The tenants of her theory are behind in their rent*). (Examples are taken from Lakoff & Johnson, 1980, pp. 98 and 110 and from Grady, 1997, p. 270.)

The partial nature of metaphorical and analogical mapping is an issue that has received considerable attention from researchers. Which entities in the source and target domains will enter into correspondences, and which particular correspondences will be set up? Several explanations have been proffered that rely on the internal structure of the spaces between which the mapping takes place.

Gentner answers the question as to which parts of the base (i.e., source) domain will be mapped to the target domain in the related study of analogy by proposing that higher-order predicates will take priority over lower-order predicates in the mapping operation. "Central to the mapping process is the principle of systematicity: People prefer to map connected *systems of relations* governed by higher-order relations with inferential import, rather than isolated predicates" (1989, p. 201). Accordingly, relationships take priority over attributes. Moreover, systems of mutually interconnecting relationships are more likely to be imported into the target domain than independent relationships (Gentner, 1983, pp. 162-163).

Lakoff offers a somewhat different account of the correspondences that are mapped onto the target domain from the source domain, referred to as the Invariance Principle or Hypothesis: "Metaphorical mappings preserve the cognitive topology (that is, the image-schema structure) of the source domain, in a way consistent with the inherent structure of the target domain" (1993, p. 215). Turner restates the hypothesis thus: "In metaphoric mapping, for those components of the source and target domains determined to be involved in the mapping, preserve the image-schematic structure of the target, and import as much image-schematic structure from the source as is consistent with that preservation" (1993, pp. 302-303). The structure inherent in the target domain thus acts as a constraint on potential correspondences between source and target domains. Turner (1993, p. 302) offers LIFE IS A JOURNEY as an example where the need to preserve the image-schematic structure of the target blocks mapping over some of the image-schematic structure of the source. In the source domain, a fork in the path is stable and is not affected by the choice made at the fork: If one returns to a traffic intersection, one expects the same choices to be available. But in life, it is often the case that when a decision has been made, at least some of the alternative options available at the time of the choice cease to be available later.

Further, in considering the issue of partial mappings, Grady, Taub, & Morgan (1996) and Grady (1997) distinguish between compound metaphors like THEORIES ARE BUILDINGS and the primitive or primary metaphors like ORGANIZATION IS PHYSICAL STRUCTURE and PERSISTING IS REMAINING ERECT of which it is composed. They characterize primary metaphors as those for which there is independent and direct experiential basis, and argue that the partiality of the mapping emerges only in the context of the compound metaphor, not in the context of the primary metaphors.

As observed, while metaphor involves mapping the internal structure of one domain onto another, the mapping is often partial. This has been explained in terms of a preference for mapping systems of relationships over the mapping of less complex components, of the inherent relations within the target domain, and of the image schemata underlying the source domain. In all cases, what is or is not mapped is constrained by the internal structure of the domains involved in the mapping.

3.3 Blended Spaces

Mental spaces are dynamically created assemblies of elements, structured by the kinds of cognitive models examined in Section 2, used in our local understanding of language and thought (Fauconnier & Turner, 1998, p. 137). Consider, for example, the sentence *When he was younger, Max believed in UFOs*. In understanding the sentence, we construct a space that is temporally removed from the present in which Max was younger than he is now; a mapping links this past Max to the present Max. There is also a belief space in which Max holds that UFOs exist in the real world, a belief he may no longer sustain. Furthermore, this UFO belief space is descended from the mental space set up for the younger Max, which is its parent space. As Sweetser and Fauconnier explain, "The basic idea is that, as we think and talk, mental spaces are set up, structured, and linked under pressure from grammar, context, and culture. The effect is to create a network of spaces through which we move as discourse unfolds" (1996, p. 11).

Blended spaces are generated by selectively projecting structure from multiple input spaces into a separate mental space (Fauconnier & Turner, 1998, p. 133; for a comprehensive overview of work in this area see Turner, n.d.). A classic example of an inferential problem-solving puzzle that relies on a blended spaces approach is the "riddle of the Buddhist monk":

One morning, exactly at sunrise, a Buddhist monk began to climb a tall mountain. The narrow path, no more than a foot or two wide, spiraled around the mountain to a glittering temple at the summit.

The monk ascended the path at varying rates of speed, stopping many times along the way to rest and to eat the dried fruit he carried with him. He reached the temple shortly before sunset. After several days of fasting and meditation he began his journey back along the same path, starting at sunrise and again walking at variable speed, with many pauses along the way. His average speed descending was, of course, greater than his average climbing speed.

Prove that there is a spot along the path that the monk will occupy on both trips at precisely the same time of day. (Gardner, 1961, pp. 168-170)

The riddle is solved by imagining the monk making both trips simultaneously, one starting at the base of the mountain at dawn and reaching the top of the mountain at sunset, and the other starting at the top of the mountain at dawn and ending at the mountain's base at sunset. Making the minimal assumption that the period between one day's dawn and its sunset overlap the period between the other day's dawn and its sunset, the monk making the trek up the mountain and the monk making the trek down the mountain must meet at some point. That point is reached at the same time going up and coming down, and is the place whose identification is sought.

Each of the journeys is set up in its own mental space, as shown in figure 3. Input space 1 pictures the monk (A1) going up the mountain on one particular day (D1), while input space 2 pictures the monk (A2) coming down the mountain on some subsequent day (D2). The path is represented schematically by a diagonal line, and a small circle represents the location of the monk on the path at some unspecified time. Figure 4 shows the correspondences between the various parts of those spaces, including links between the (opposite) directions in which travel is taking place, the base and top of the mountain (or alternatively, the endpoints of the path being traversed), the days on which travel takes place, and the monk's location on the path. Finally, figure 5 selectively projects parts of the structure of the two input spaces into a blended space, in which the two days, D1 and D2, become a unified day, and in which the monk, in his travels on the path, becomes two simultaneous travelers, one ascending and the other descending.

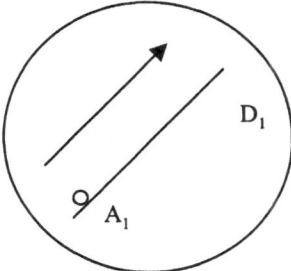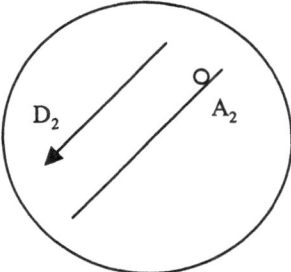

Figure 3. Original input spaces

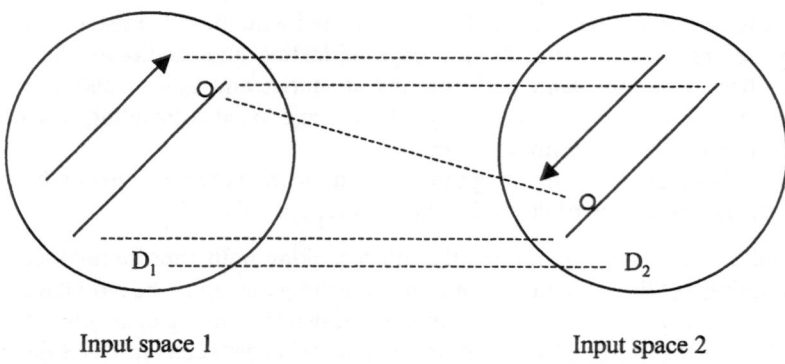

Input space 1   Input space 2

Figure 4. Cross-space mapping of input spaces

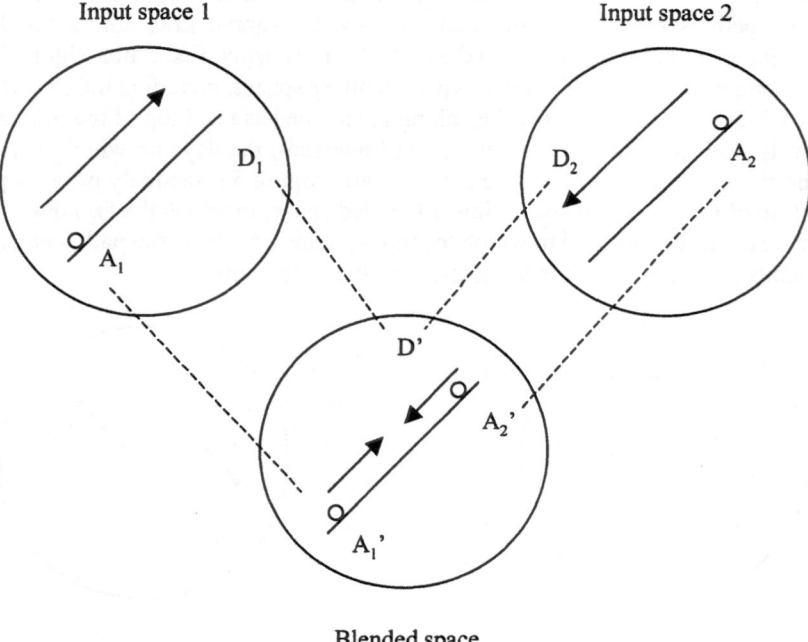

Blended space

Figure 5. Mapping from input spaces to blended spaces

Not only do we find the basic schematic structure of the input spaces repeated in the blended space, with relationships noted as correspondences between the elements of the input spaces and their counterparts in the blended space, but we also find emergent structure in the blended space. Here, and only here, are there two travelers on the path on the same day, moving in opposite directions, whose relative location can be compared (Fauconnier & Turner, 1998, pp. 136-140).

Analogical counterfactuals also admit of a blended spaces representation. Examples of counterfactuals include the following (taken from Fauconnier, 1994, 1996):

In France, Watergate wouldn't have done Nixon any harm.
Coming home, I drove into the wrong house and collided with a tree I don't have.
If I had been Reagan, I wouldn't have sold arms to Iran.
If Napoleon had been the son of Alexander, he would have won the battle of Waterloo.
If Woody Allen had been born twins, they would have been sorry for each other.

Our understanding of such concepts relies on merging corresponding components of a factual space and a counterfactual space into a single blended space. Again, the internal structure of those spaces is required before correspondences can emerge.

4. CONCLUSION

The basic conceptual units of cognitive semantics—image schemata, basic level categories, and frames—are structured. This structure arises from relationships that exist between components of the conceptual unit. In the case of the prototypical basic level category, these components are parts of physical entities; in the case of image schemata and frames, these components are roles within the conceptual unit.

Mappings within and between domains/frames create further conceptual structure. Metonymic mappings depend on specific types of relationships within frames, while metaphor and conceptual integration rely on projecting structure from one or more base models onto a target domain or blended space.

As has been shown, structured conceptual models organize our conceptual world at seemingly every turn. Simply put, the primary components of thought and language cannot be comprehended without recourse to structure and relationships.

References

Barsalou, L. W. (1992). Frames, concepts, and conceptual fields. In A. Lehrer & E. F. Kittay (Eds.), *Frames, Fields, and Contrasts: New Essays in Semantic and Lexical Organization,* 21-74. Hillsdale, NJ: Erlbaum.
Brown, R. (1958). How shall a thing be called? *Psychological Review,* 65, 14-21.
Clausner, T., & Croft, W. (1999). Domains and image schemas. *Cognitive Linguistics,* 10, 1-31.
Fauconnier, G. (1994). *Mental Spaces: Aspects of Meaning Construction in Natural Language* (2nd ed.). Cambridge: Cambridge University Press.

88   R. Green

Fauconnier, G. (1996). Analogical counterfactuals. In G. Fauconnier & E. Sweetser (Eds.), *Spaces, Worlds, and Grammar,* 57-90. Chicago: University of Chicago Press.

Fauconnier, G. (1997). *Mappings in Thought and Language.* Cambridge: Cambridge University Press.

Fauconnier, G., & Turner, M. (1998). Conceptual integration networks. *Cognitive Science,* 22, 133-187.

Fillmore, C. J. (1976). Frame semantics and the nature of language. In S. Harnard, H. Steklis, & J. Lancaster (Eds.), *Origins and Evolution of Language and Speech,* 20-32. New York: New York Academy of Science.

Fillmore, C. J. (1982). Frame semantics. In *Linguistics in the Morning Calm,* 111-137. Seoul: Hanshin.

Fillmore, C. J., & Atkins, B. T. (1992). Toward a frame-based lexicon: The semantics of RISK and its neighbors. In A. Lehrer & E. F. Kittay (Eds.), *Frames, Fields, and Contrasts: New Essays in Semantic and Lexical Organization,* 75-102. Hillsdale, NJ: Erlbaum.

FrameNet. (n.d.). Available: <http://www.icsi.berkeley.edu/~framenet/> [2001, October 9].

Gardner, M. (1961, June). Mathematical games. *Scientific American,* 204, 166-176.

Gentner, D. (1983). Structure-mapping: A theoretical framework for analogy. *Cognitive Science,* 7, 155-170.

Gentner, D. (1989). The mechanisms of analogical learning. In S. Vosniadou & A. Ortony (Eds.), *Similarity and Analogical Reasoning,* 199-241. Cambridge: Cambridge University Press.

Grady, J. E. (1997). THEORIES ARE BUILDINGS revisited. *Cognitive Linguistics,* 8, 267-290.

Grady, J., Taub, S., & Morgan, P. (1996). Primitive and compound metaphors. In Goldberg, A. E. (Ed.), *Conceptual Structure, Discourse and Language,* 177-187. Stanford, CA: Center for the Study of Language and Information.

Johnson, M. (1987). *The Body in the Mind: The Bodily Basis of Meaning, Imagination, and Reason.* Chicago: University of Chicago Press.

Kövecses, Z., & Radden, G. (1998). Metonymy: Developing a cognitive linguistic view. *Cognitive Linguistics,* 9, 37-77.

Krzeszowski, T. P. (1993). The axiological parameter in preconceptional image schemata. In R. A. Geiger & B. Rudzka-Ostyn (Eds.), *Conceptualizations and Mental Processing in Language,* 307-329. Berlin: Mouton de Gruyter.

Lakoff, G. (1987). *Women, Fire, and Dangerous Things: What Categories Reveal about the Mind.* Chicago: University of Chicago Press.

Lakoff, G. (1993). The contemporary theory of metaphor. In A. Ortony (Ed.), *Metaphor and Thought* (2nd ed.), 202-251. Cambridge: Cambridge University Press.

Lakoff, G. (1994). Conceptual metaphor home page. Available: <http://cogsci.berkeley.edu> [2001, October 9].

Lakoff, G., & Johnson, M. (1980). *Metaphors We Live By.* Chicago: University of Chicago Press.

Langacker, R. W. (1987). *Foundations of Cognitive Grammar* (Volume I: *Theoretical Prerequisites).* Stanford: Stanford University Press.

Langacker, R. W. (1991). *Concept, Image, and Symbol: The Cognitive Basis of Grammar.* Berlin: Mouton de Gruyter.

Lindner, S. (1981). *A Lexico-Semantic Analysis of Verb-Particle Constructions with UP and OUT.* Doctoral dissertation, University of California, San Diego.

Lindner, S. (1982). What goes up doesn't necessarily come down: The ins and outs of opposites. *Papers from the Regional Meeting of the Chicago Linguistic Society*, 18, 305-323.

Mill, J. S. (1973). *A System of Logic, Ratiocinative and Inductive: Being a Connected View of the Principles of Evidence and the Methods of Scientific Investigation* (J. M. Robson, Ed.). Toronto: University of Toronto Press. (Original work published 1843).

Rosch, E., Mervis, C., Gray, W., Johnson, D., & Boyes-Braem, P. (1976). Basic objects in natural categories. *Cognitive Psychology*, 8, 382-439.

Sweetser, E., & Fauconnier, G. (1996). Cognitive links and domains: Basic aspects of mental space theory. In G. Fauconnier & E. Sweetser (Eds.), *Spaces, Worlds, and Grammar,* 1-28. Chicago: University of Chicago Press.

Turner, M. (1993). An image-schematic constraint on metaphor. In R. A. Geiger & B. Rudzka-Ostyn (Eds.), *Conceptualizations and Mental Processing in Language,* 291-306. Berlin: Mouton de Gruyter.

Turner, M. (n.d.). Blending and conceptual integration. Available: <http://www.wam.umd.edu/~mturn/WWW/blending.html> [2001, October 9].

Tversky, B. (1986). Components and categorization. In C. Craig (Ed.), *Categorization and Noun Classification*, 63-76. Philadelphia: Benjamins.

Tversky, B., & Hemenway, K. (1984). Objects, parts, and categories. *Journal of Experimental Psychology: General*, 113, 169-193.

Chapter 6

Comparing Sets of Semantic Relations in Ontologies

Eduard Hovy
Information Sciences Institute, University of Southern California, Marina del Rey, CA, USA

Abstract:
　　A set of semantic relations is created every time a domain modeler wants to solve some complex problem computationally. These relations are usually organized into ontologies. But there is little standardization of ontologies today, and almost no discussion on ways of comparing relations, of determining a general approach to creating relations, or of modeling in general. This chapter outlines an approach to establishing a general methodology for comparing and justifying sets of relations (and ontologies in general). It first provides several dozen characteristics of ontologies, organized into three taxonomies of increasingly detailed features, by which many essential characteristics of ontologies can be described. These features enable one to compare ontologies at a general level, without studying every concept they contain. But sometimes it is necessary to make detailed comparisons of content. The chapter then illustrates one method for determining salient points for comparison, using algorithms that semi-automatically identify similarities and differences between ontologies.

1. INTRODUCTION

Over the past decades, semantic relations have been the focus of much work by philosophers, computer scientists, linguists, and others. They have developed formalisms, built computational reasoning systems, collected sets of relations, and arranged taxonomies of relations or ontologies. Thus, even for describing the same domain, it is not surprising that many different and mutually incompatible systems of semantic relations exist!

Certainly no set of relations is entirely complete, entirely right, or entirely wrong. To study their collected insight, one might then try to compare them, drawing from each that which is valuable for one's enterprise. But despite some discussion of the relative merits of the various approaches and results, there have been few proposals for empirical ways in which to compare and justify them. At present, knowledge crafters follow their own paths and derive their own results, without much regard to standardized procedures and/or methods of comparing their work to that of others.

Surely however it is not impossible to establish some methodology and/or standards for description and comparison. One could begin by investigating typical methods of creating, justifying, and validating sets of relations and ontologies in general, and then identify the primary commonalities and points of importance. From that, one could derive a list of the kinds of characteristics that are important when one describes an ontology or set of relations, and when one compares various ontologies or sets of relations to one another.

This chapter lays some foundations for doing so, focusing on ontologies as sets of relations and concepts. For generality, we define an ontology rather loosely as a set of terms, associated with definitions in natural language (say, English) and, if possible, using formal relations and constraints, about some domain of interest, used in their work by humans, data bases, and computer programs. We view a set of semantic relations, organized into collections and perhaps related in a generalization hierarchy, as a special instance of an ontology. We view a domain model as an ontology that specializes on a particular domain of interest.

The steady emergence of new and different ontologies is making it possible to perform ontology characterization and comparison studies. In fact, one can compare at two levels: using the overall general characteristics of the ontologies and focusing on particular differences in the ontologies' content and structure. The former level requires a framework of characteristics that covers all pertinent aspects, in terms of which one can then characterize each ontology. The latter level requires a method of identifying individual points of difference, for which one can then study differences between specific relations, concept definitions, and interrelatedness with other concepts and/or relations.

To compare ontologies, one can proceed as follows. Given two ontologies or sets of relations, one can first create the general characterizations for each, then identify overall differences, and finally identify particular points of difference between individual pairs of concepts or relations.

This chapter describes the first steps in performing this complex undertaking. Section 2 provides an overall framework of characteristics of ontologies in general, and Section 3 describes a method for identifying individual concept-by-concept differences between ontologies, using semi-automated alignment algorithms.

2.   CHARACTERIZING AN ONTOLOGY

2.1   The Need to Characterize Ontologies

The first step outlined above is to characterize an ontology in terms of a general set of descriptive features. To date, only a few systems of features for describing or comparing ontologies have been proposed (Noy & Hafner, 1997; Uschold, Clark, Healy, Williamson, & Woods 1998; Aguado et al., 1998; Jasper & Uschold, 1999; Van Zyl & Corbett, 2000). The framework described in this section was first disseminated at a American Association for Artificial Intelligence (AAAI) Symposium in March 1997 and is offered as a stepping stone on the way toward a fuller and richer set of features. There is no claim that the features outlined here, or their organization, is the only one, or that it is the best possible one. These ideas are meant to suggest a possible avenue of attack; the major claims lie in the approach, and its minor claims in the set of features chosen at each level.

The descriptive features given here are supposed to capture the salient aspects of ontologies in a single standardized way, in order to facilitate cross-ontology comparisons (and possibly, eventually, ontology evaluation). This system recognizes our desire for simple, intuitively clear, approachable features while respecting the complexity and nonconformity of ontologies we encounter today. To allow this, the features are organized into a taxonomy of increasing differentiation and specificity, thereby bounding (on the horizontal axis) the number of (sub)cases to examine at any point, affording simplicity, while still allowing unbounded specialization on the vertical axis, downward, accommodating variability. (Eventually producing an ontology of ontologies would be a most satisfying result.)

The taxonomy of features is described top-down, for clarity of exposition. Unfortunately, this makes term definition difficult, because many terms are most easily explained simply by considering the subtaxonomy they dominate. Striking a compromise, each level includes a fairly informal, intuitive, definition of the terms involved, trusting that the reader will keep dissatisfaction in check until the subsequent level down.

2.2 A System of Features—Upper Levels

At the most fundamental level of description, one has to distinguish the most basic sets of concerns. For ontology engineering (as for knowledge representation [KR] in general), a useful top-level division separates form, content, and usage. In this section we provide a higher-level overview and then, in Section 2.3, provide specifics.

2.2.1 Form

Form denotes the representational framework—the conceptual tools with which the ontology builder defines terms and axioms. Providing the representational substrate, it includes theoretical, notational, and computational concerns. Mainly, this area involves issues in mathematical logic and computational complexity. Form is usefully separated into two branches:

- *Theoretical foundation* includes four subareas, namely the conceptual roots of the ontology, the principles of terminology design, the denotational theory used, and the treatment of microtheories (contexts).
- *Computational foundation* includes two principal subareas, namely the properties and capabilities of the KR system employed and the notation or formalism used.

These two notions may be partitioned as follows:

- *Conceptual roots* pertains to the historical antecedents of the ontology, including all the ontologies and KR systems upon which the current ontology is based and from which design decisions have been drawn.
- *Principles of terminology design* includes all the considerations that lead to the ontology builders' selection of terms to model aspects of the domain, such as how parsimonious to be, and on what basis to taxonomize the terms.
- *Denotational theory* denotes the theory/ies used to establish the formal link between the ontology and its extension in the world of the domain; the approach might be formal, such as possible worlds semantics, or it might be informal, relying on natural language glosses.

- *Microtheories or contexts* pertains to the creation and treatment of subareas of the ontology, or collections of axioms, that hold only in particular circumstances; for example, animals can talk only in children's stories. This aspect contains a description of the particular features of microtheories or contexts required in the ontology.
- *Properties and capabilities of KR system* includes all aspects of the representation and reasoning support required by the ontology of the underlying knowledge representation system, from simple capabilities such as property inheritance to complex ones such as truth maintenance.
- *Notation (formalism)* denotes the design of the notation used, both for defining the ontology's concepts and relations, and for representing instances in the domain by composing terms from the ontology.

The final level of detail for Form is discussed in Section 2.3.1.

2.2.2 Content

Content denotes the terms, axioms, and microtheories used by the ontology builder to model the domain of interest. Mainly, this area involves issues in philosophy, epistemology, and, naturally, the domain itself. Content is usefully separated into four branches: terminology, axioms, inferences, and instances.

- *Terminology / feature organization* pertains to the collection of terms (relations, objects, events, etc.) taken as a whole, as included in the ontology by the ontology builder; important aspects include the number of terms, organization, coverage over domain, etc. This topic may even include the terms themselves.
- *Microtheories / contexts* describes pertinent aspects of the different microtheories that have been defined in and for the ontology under consideration.
- *Inferences* includes all aspects of the inferences defined over the terms in the ontology (to the extent this is separate from the axioms). It may also include the actual inferences themselves.
- *Individual terms or instances* pertains to the characteristics of the ontology's representation items, taken individually. This includes the typical quantity and quality of content associated with each term, the expressive adequacy of terms when it comes to representing the domain(s) of interest, the compositionality of terms, and so on.
- *Axioms* includes all aspects of the domain axioms and other interrelationships defined among the terms in the ontology. It may also include the actual axioms and relations themselves.
- *Conceptual roots* describes the sources of the ontology content, its development history, and methods of extracting and/or creating the content.

The final level of detail for Content is discussed in Section 2.3.2.

2.2.3 Usage

Usage denotes the manner in which the ontology builder goes about building, using, updating, and maintaining the ontology, including the software support and tools for these activities. It involves issues in software engineering and ergonomics. Usage can be subdivided into the following four areas:

- *Functionality* pertains to all aspects of the actual and intended use(s) of the ontology, mainly in implemented systems, but also in its historical perspective, exemplifying, for instance, what has been achieved with it.
- *Acquisition* describes the legal and administrative details of acquiring the ontology.
- *Implementation* denotes relevant aspects of the computational instantiation(s) of the ontology, such as storage size, inference speed, hardware and software required, etc.
- *Support* includes all aspects of ontology builder and user support, such as viewing and editing tools, error handling, documentation, etc.

The final level of detail for Usage is discussed in Section 2.3.3. It is interesting to note that the knowledge representation field ignored usage as a topic of theoretical interest almost from its inception—papers discussing implementations, inference speed, etc., appear in more applied conferences only. Similarly, it factored out, and mostly ignored, issues of content a little later, certainly by the time of Brachman's (1978) influential KL-ONE system.

2.3 Features—Detail Level

2.3.1 Features of Form

An outline of the features of form detailed below is given in figure 1.

The conceptual roots include all the ontologies and KR systems upon which the current ontology is based and from which design decisions and content have been drawn. Knowing its derivation (and the beliefs of its builders) may help one understand the ontology itself.

The principles of terminology design include all the considerations that lead to the ontology builders' selection of terms to model aspects of the domain:

- The *degree of terminology parsimoniousness*: This feature characterizes the number of the ontology terms, on the scale from parsimonious (very few terms, all 'primitive') to profligate (very many terms, some quite specific). For example, a very parsimonious term set, Conceptual Dependency, contained only 14 action primitives (Schank & Abelson, 1977). In contrast, an argument in favor of profligacy can be found in the paper *Ontological Promiscuity* (Hobbs, 1985). This feature involves what might be called the *principle of term inclusion*; any careful ontology builder should be able to state the general policy being followed regarding the granularity of terms and the resulting effect on the ontology size. Also to be described is the extensibility of the term set—the ease of adding new terms, both

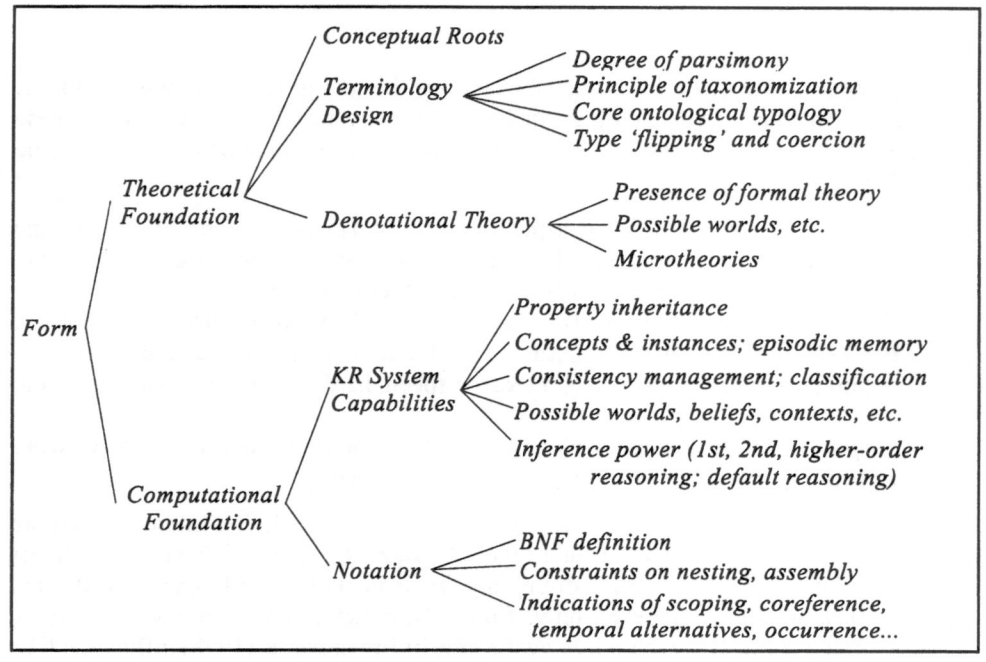

Figure 1. Features of form

defining them adequately to the level of the ontology and finding their proper
location(s) in the ontology.

- One or more *principle(s) of taxonomization*, setting forth the basis from which the
  ontology's taxonomic / network structure is derived, and by which the placement
  of new terms is justified and tested. In the Pangloss Ontology Base, for example,
  the principle is derived from the English grammar encoded in the Penman sentence
  generator (Bateman, Kasper, Moore, & Whitney, 1989); in WordNet, the principle
  appears to be based on what might be called *naive semantics* with a cognitive
  science bent (Fellbaum, 1998); in the CYC ontology, the principle seems to be
  based on intuitions guided by artificial intelligence inference concerns (Lenat,
  1995).
- The *core ontological typology*, namely the major ontological types, and the theory
  of the domain that underlies them. Included here are questions about states vs.
  events vs. state-changes vs. processes; relations as states or not as states; the
  relationships between object and event (see 'type flipping' below). A popular
  approach (taken, for example, by Aristotle), starts with a given list of the basic
  types that function as the roots of the forest of trees forming the ontology. Another
  approach is to identify a set of very basic facets of meaning (e.g., concrete-abstract,
  positive-negative, changing-static) whose various permutations form the starting
  point for ontological differentiation (Wille, 1997; Sowa, 1999). Despite many

interesting theories about the basic ontological types, little formal agreement exists, although a fairly standard general practical approach seems to be emerging.

*   The *standard for 'type flipping' and coercion*: Regardless of particular decisions on the core ontological typology, it seems to be a fact about the world and the way people model it that certain concepts receive treatment sometimes as one basic type and sometimes as another. Often the concept is sufficiently different in its alternate guise to merit being called a distinct (but still somehow related) concept. Various classes of 'type flipping' can be differentiated. In metonymy, an invalid type is substituted for another (for example, "In a press statement, the Senate announced that . . .", or "Last year, London attracted 10 million visitors"—here *Senate* stands for a spokesperson, since the building obviously cannot make announcements, and *London* for the social organization that is the city, not its physical location, mayoral office, etc.). In type promotion and demotion, some portion of a representation is moved to a different position with respect to the whole, possibly causing changes in terms used (for example, "Mike swims across the river" may initially be represented as a *swim* process but then be changed to a *move* or even a *move-across* process whose manner is *swimmingly*—this kind of change is common in shallow semantic representations in machine translation when different languages express different aspects of the same idea (Dorr, 1994)). Though such type flipping is readily apparent in linguistic applications, it appears also in artificial intelligence (AI) reasoning. For example, some concepts can be treated as either objects or processes (e.g., are *picnic, parade,* and *explosion* events or objects? How do *explosion* and *to explode* relate? Is a *street corner* a location or an object? And a *city*? Ontology builders should state their policy on the treatment of type coercion, since this affects their approach to ontology creation.

The denotational theory denotes the theory/ies used to establish a formal link between the ontology and its extension in the world of the domain. Several aspects are important, including:

*   *Presence of logically rigorous denotational system*: Possible alternatives range from 'informal' (i.e., without a logically rigorous denotational system) to formal, with the former used more in implemented systems and the latter developed mainly by (and for?) theoreticians. In informal approaches, the world-extension of a representational item is generally simply inferred from its name, which is a carefully chosen natural language word; the dangers of this approach were tellingly pointed out a long time ago by McDermott (1981). Formal approaches include work by Montague (1974), Davidson (1967), Barwise and Perry (1986) on situational semantics, and others.
*   The possibilities for representing distinct *contexts or worlds* of various kinds, including possible worlds.
*   *Microtheories*, which provide the facility of encapsulating a set of terms, relations, and/or axioms into a system distinct from the rest of the ontology, and applied to the world under specific conditions only (for example, in childrens' stories, animals can talk). The CYC ontology supports microtheories for inference (Lenat, 1995); MIKROKOSMOS for machine translation (Mahesh, 1996; Nirenburg, Raskin, & Onyshkevych, 1995).

The properties and capabilities of the KR system include the representation and reasoning support required by the ontology of the underlying knowledge representation system. It is obviously closely related to the theoretical foundation. Since this area is one of the mainstays of KR research, a large number of such properties and capabilities have been discussed and implemented, including:

- The treatment of *property inheritance*, whether or not it is supported, whether it is strict (i.e., no violations allowed) or not (i.e., overrides allowed), and how the Nixon diamond and similar problems are handled;
- The *differentiation of concepts and instances*, which are separated in KL-ONE and successors like Loom (MacGregor, 1991) and CLASSIC (Borgida, Brachman, McGuinness, & Alperin Resnick, 1989; Brachman, Borgida, McGuinness, & Patel-Schneider, 1999) into the T-Box and A-Box, but not so separated in for example CYC or the name taxonomies of the Yahoo! and Lycos internet search engines;
- The *support of inference,* which involves both the types supported (active/forward inference, passive/backward inference, or none), and includes the presence and necessity of additional capabilities such an automated concept classifier (MacGregor, 1991);
- The possibilities for representing *contexts or worlds* of various kinds, including possible worlds, microtheories, etc., with details;
- The necessity of or support for a *truth maintenance* system, and its details;
- The necessity of or support for a *time management* system, and its details;
- The necessity of or support for various forms of *non-first-order reasoning,* including second- and higher order and modal reasoning;
etc.

The notation (formalism) denotes the design of the notation used to represent instances in the domain by composing terms from the ontology. Many alternatives, often simply notational variants that differ in the placement of parentheses and other markings, have been developed; two major families are the frame-style representations and the axiom-style ones. Several characteristics must be described, including:

- A *BNF description* of the notation;
- Any constraints on the *nesting and assembly* of representation instances, including a description of equivalent variant forms, if the notation allows this;
- The notations used for *indicating various special situations*, including scoping, coreferences (repeated occurrences of the same entity), temporal alternatives (the same entity at different times), actual vs. potential occurrences of events in the domain, and so on;
etc.

2.3.2 Features of Content

An outline of the features of content detailed below is given in figure 2.
The terminology and feature organization includes all aspects of the actual terms included in the ontology, among others:

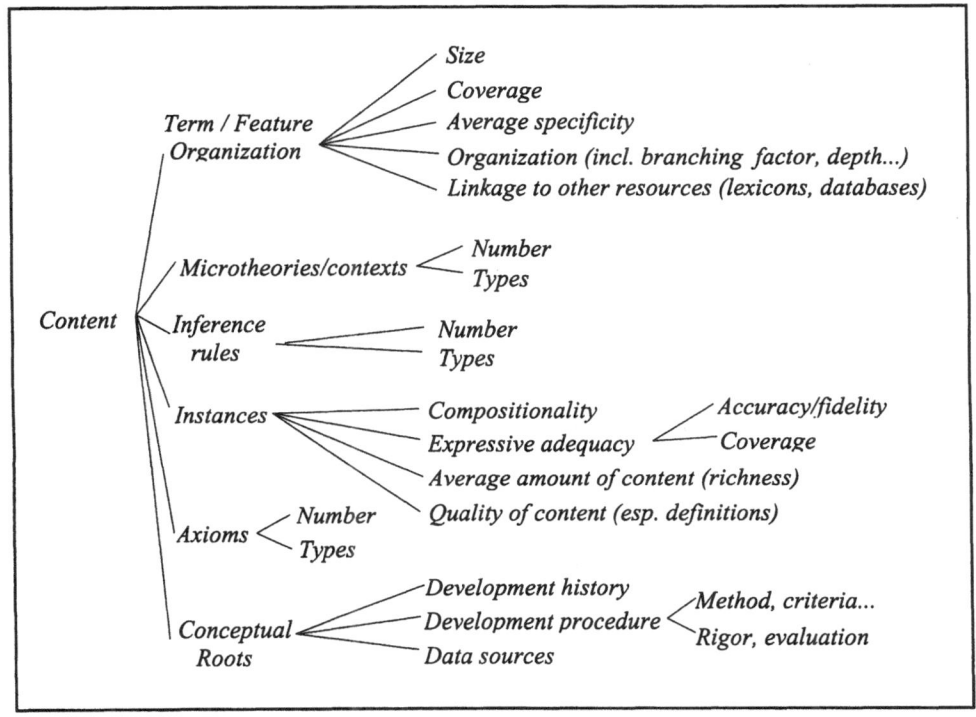

Figure 2. Features of content

- *Size* of the ontology is expressed by the number of its terms.
- *Coverage* over the domain refers to the percentage of concepts identified in the domain that are explicitly represented by terms in the ontology.
- *Richness* of terminology is measured in various ways (for example, number of predications per term, or total number of interconnections among terms). Terminological ontologies, such as SENSUS (Knight & Luk, 1994), contain mostly terms organized hierarchically and are less rich (but usually much larger) than domain models.
- *Organization* of terminology may be a simple tree taxonomy or an arbitrarily complex network. If it is a simple taxonomy, or a set of parallel taxonomizations over the same terms, then the taxonomizing principle(s) must be described (see under *terminology design* above). Often, the organizing principle is some variant of conceptual generalization, using *is-a* or *subclass* and *instance-of;* it might also be *part-of, synonym-to, antonym-to,* etc., as in WordNet (Fellbaum, 1998). Various subsidiary measures provide additional information, including the average branching factor, the average depth of each major portion of the ontology, etc.
- Relative to the average level of *specificity* of the terms, terminological ontologies used for natural language applications tend to be more general (high-level, abstract), especially such language-related ontologies as the Penman Upper Model

(Bateman Kasper, Moore, & Whitney, 1989), while domain models used for domain-oriented applications are naturally more specific.

- *Linkages*, if any, go from the terms into other resources, such as lexicons of various languages (as typically used in machine translation (Knight & Luk, 1994)), termlists acquired from outside agencies (often present in ontologies modeling database contents), etc.
- Lastly is the set of *terms* themselves.

The microtheories that included the ontology—their *number* and *types*—are listed and described.

Domain inference rules include all aspects of the inferences defined over the terms in the ontology (to the extent that this is separate from the axioms), including:

- The *number* of inference rules defined;
- The *inferential capabilities* required or permitted of the inference rules: whether, for example, inferences may create new instances, delete them, add properties and other relations, change their status from (for example) actual to potential, etc.; and
- The actual *inferences* themselves.

Instances comprise the aspects of representational instances created for actual domain entities existing at specific times with states of being. This includes the following:

- *Expressive adequacy*, which involves both the *fidelity* or correctness of the instance (that is, the degree to which it represents what the ontology builder or system user believes to be true about the actual instance in the domain—*the truth, and nothing but the truth*), and the *coverage* or completeness of the instance (that is, the degree to which it represents all the facts about the instance in the domain—*the whole truth*);
- *Compositionality*, which includes the ease of composing instances, either anew or out of other instances, the readability of instances, ease of access to their parts, etc.; etc.

The axioms of the ontology, including any other interrelationships defined among the terms, involve the following aspects:

- The *number* of axioms and/or relationships defined;
- The *types* of axioms and relations allowed in the ontology;
- The actual *axioms and relations* themselves.

Conceptual roots lists the *sources* of all the content terms themselves and describes their *history of development* as well as the *procedures* by which they were co-opted, altered, merged, etc.

2.3.3 Features of Usage

An outline of the features of usage detailed below is given in figure 3.

Functionality pertains to all aspects of the actual and intended use(s) of the ontology, mainly in implemented systems, but also in its historical perspective, exemplifying, for

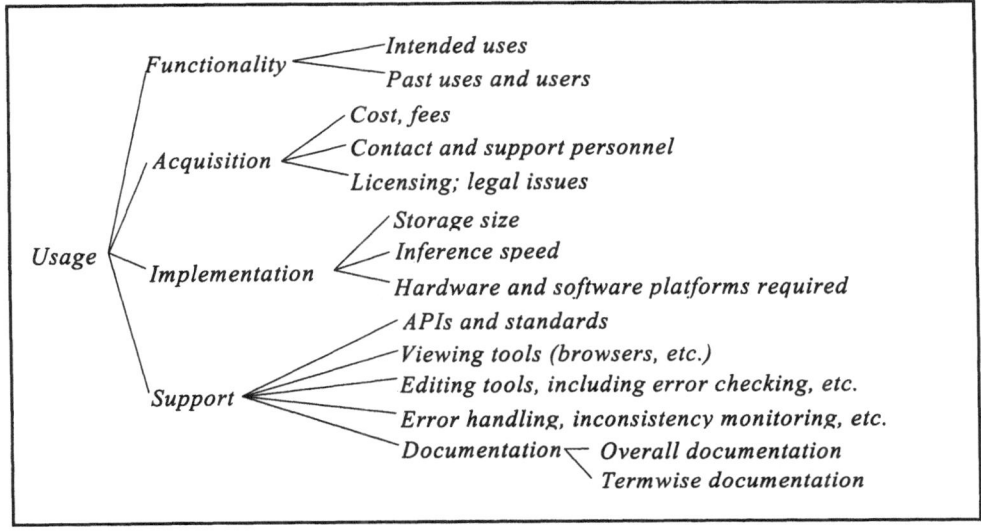

Figure 3. Features of usage

instance, what has been achieved with it. Listed here are both the *intended uses*, which can play a major part in one's attitude toward the utility of an ontology (see Van Zyl & Corbett, 2000 and Ushold, Clark, Healy, Williamson, & Woods, 1998, for example), and *past and present users*, namely those to whom one can turn for further information, experience, etc.

Acquisition describes the legal and administrative details of acquiring the ontology. This includes *licensing, cost*, etc.

Implementation denotes relevant aspects of the computational instantiatation(s) of the ontology. This includes:

- *Storage size*, both of the ontology when stored on disk and when loaded in core;
- *Inference speed* and speed of other computations on the ontology;
- Hardware and software *platforms required* by the ontology's KR system(s); etc.

Support includes all aspects of computational support for the ontology builder and user, including:

- *APIs and standards*: a description of the degree and rigor of standardization of the ontology as data;
- Viewing tools: the presence and quality of tools such as browsers, ontology prettyprinters, Web-based interfaces, etc.;
- *Editing tools*, with or without version control, automated error checking, etc.;
- *Error handling*, including inconsistency monitoring;
- *Documentation*: its presence and quality; etc.

3. COMPARING CONCEPTS OR RELATIONS SEMI-AUTOMATICALLY

Comparing two sets of relations or two ontologies using the framework just described, one can create a list of some global differences between them. Often, however, it is most useful to frame one's questions in terms of specific point-by-point differences between individual concepts or relations. When one has identified alternative (equivalent, or somewhat equivalent) concepts or relations, one can use especially the features of content to understand the details of variation.

However, when presented with two ontologies or sets of relations, each containing several thousand items, how does one find the (near-)equivalent ones? One cannot manually compare each concept in one ontology to all other concepts in the other; this approach may require $M \times N$ comparisons (for M and N items respectively in the two ontologies); if each comparison takes just one minute, two 5,000-item ontologies would require over 200 person-years. An automated search is clearly required, if only to pinpoint likely candidate equivalent items. But even if each concept or relation is fully specified by logical propositions, what search mechanism can be conceived to 'understand' the definitions well enough?

In recent work, several ontology building projects have build interfaces to assist with the manual alignment and merging of ontologies. Typically, these tools extend ontology building interfaces such as those of Ontolingua (Farquhar, Fikes, & Rice, 1997) and the Stanford CML (Compositional Modeling Language) Editor (Iwasaki, Fraquhar, Fikes, & Rice, 1997) by incorporating variants of matching heuristics plus several validation routines that check for consistency of edited results (McGuinness, Fikes, Rice, & Wilder, 2000; Noy & Musen, 1999a, 1999b). Given the large numbers involved, however, we are interested in automating the alignment process.

This section outlines one experiment in automatically identifying potentially (near-) equivalent concepts or relations, using several techniques which each provide contributory evidence, which evidence can then narrow down the amount of comparative work a human analyst has to perform. By applying this method and discarding invalid suggestions, the analyst can more quickly and easily produce a list of purportedly equivalent concepts or relations, whose differences can then be studied one by one. Section 3.1 outlines the context of cross-ontology comparison work; Section 3.2 describes the alignment technique, and Section 3.3 provides some results.

3.1 Background: Searching for a Standard Reference Ontology

In a world of increasing computerization, it is inevitable that different computer programs represent the same basic entity in different ways. The concept *Person*, for example, connotes something different to a medical specialist (and hence to his or her computer program) than it does to a Census Bureau worker (and his or her program) or a video game player. But in many of these applications, much of the information associated with the entity is the same. To the extent that establishing a single, large, standard ontology of terms is feasible, it would hold at least the following advantages: standardization of terms, easier knowledge transfer to new domains, and better interoperability between computer programs.

Recognizing these advantages, members of several ontology projects and others interested in the issues formed an Ad Hoc Group on Ontology standards in 1996. Meeting roughly twice a year, the group became a subcommittee of the ANSI Committee X3T2, later called the National Committee for Information Technology Standards (NCITS). It included representatives from various universities (including Stanford University and the Natural Language Group at the University of Southern California's Information Sciences Institute [USC/ISI], research laboratories (including LADSEB [the Institute for Systems Science and Biomedical Engineering] in Italy and Lawrence Berkeley Laboratories in California), companies (including EDR [Electronic Dictionary Research] in Tokyo, CYC in Texas, IBM in California, and TextWise in New York), U.S. Government officials, and private individuals.

The group sought to create a short document to serve as a model ontology standard and to create a small illustrative example of such a standard, called the Reference Ontology. As described in Section 3.2, the group commissioned the author to create early versions of the topmost portions of this Reference Ontology out of the top regions of the SENSUS, MIKROKOSMOS, and CYC ontologies. Although no Reference Ontology exists today, the general technique and the heuristics are of interest for this chapter.

SENSUS (Knight & Luk, 1994) was built at USC/ISI. SENSUS currently contains approx. 90,000 terms, linked together into a subsumption (*is-a*) network, with additional links for part-of, pertains-to, and so on. SENSUS is a rearrangement and extension of WordNet (Fellbaum, 1998) (built at Princeton University on general cognitive principles), retaxonomized under the Penman Upper Model (Bateman, Kasper, Moore, & Whitney, 1989) (built at USC/ISI to support natural language processing). For most of its content, SENSUS is identical to WordNet 1.5; its primary structuring is hence based on cognitive grounds. It can be characterized as a 'shallow', lexically oriented, term taxonomy.

SENSUS can be accessed publicly via the Web, using the Ontosaurus browser (Swartout, Patil, Knight, & Russ, 1996), at *http://mozart.isi.edu:8003/sensus/sensus_frame.html*. SENSUS has been used to serve as the internal mapping structure (the Interlingua termbank) between lexicons of Japanese, Arabic, Spanish, and English, in several projects, including the GAZELLE machine translation engine (Knight et al., 1995) and the SUMMARIST multilingual text summarizer (Hovy & Lin, 1999). The lexicons contain over 120,000 root words in Japanese, 60,000 in Arabic, 40,000 in Spanish, and 90,000 in both English and Bahasa Indonesian, of which various amounts have been linked to SENSUS. SENSUS terms serve as connection points between equivalent language-based words.

MIKROKOSMOS (Mahesh, 1996), built at the Computing Research Laboratory of the New Mexico State University (NMSU), contains 4790 concepts. MIKROKOSMOS was designed to support interlingual machine translation between a variety of languages. The interlingua definition symbols are housed in MIKROKOSMOS, together with internal interrelationships and property constraints. MIKROKOSMOS is thus somewhat 'richer' (in the sense of Section 2.3.2) than SENSUS, and its primary structuring is oriented toward supporting the linguistic generalizations that support language processing. Its concepts, also taxonomized under the *is-a* relation, partially resemble the upper portions of SENSUS, although they contain many additional abstractions. Many of the MIKROKOSMOS concepts are not lexicalizable in English by a single word.

The CYC Ontology (Lenat, 1995) has been under development by CYCorp Inc., Texas, since the mid-1980s, in an ambitious attempt to create a single large ontology in terms of which one can model the whole world. At present, the full CYC ontology contains about 40,000 concepts and over 300,000 axioms (inter-concept relations and constraints), and is used primarily to support metadata schemas for large pharmaceutical databases. CYC's concepts are organized according to two relations, *isa* and *genls*, that express two different types of subsumption (essentially *element-of* and *subset-of*); this approach avoids many problems experienced by less sophisticated ontologies. The CYC ontology is intended to support general conceptual inference for artificial intelligence systems; it contains semantic models of time, space, matter, and other experiential phenomena. It is not oriented toward supporting language-based applications; the majority of its concepts are not lexicalizable by single English words. There is no stated theory of semantics or systematic methodology of semantic modeling. Still, the CYC ontology is the 'richest', most semantic, of the three ontologies discussed here. Currently available to the public are only the topmost 2,500 concepts of the CYC ontology (see http://www.cyc.com), but no inter-concept relations other than the two for subsumption.

3.2 Aligning Ontology Terms

To fuse together the topmost regions of CYC (2500 concepts) and SENSUS (350 concepts) in 1996, several cross-ontology alignment and validation heuristics, described in this section, were developed and used (Hovy, 1996; Lehmann, 1997; Chalupsky, 1996). In 1997, the author extended the heuristics to also align MIKROKOSMOS (4800 concepts) to the topmost region of SENSUS (6768 concepts) (see Hovy, 1998).

The technique used combines the heuristics in a repeated alignment/integration cycle, which employs:

- Three heuristics (NAME match, DEFINITION match, and TAXONOMY match) that make initial cross-ontology alignment suggestions,[1]
- A function for combining their suggestions,
- Several alignment validation criteria and heuristics,
- An evaluation metric.

Despite the apparent difficulty of having machines 'understand' concepts well enough to suggest alignments, the heuristics described here perform well enough to identify a workable number of concept pairs.

In aligning the MIKROKOSMOS ontology with the top region of SENSUS, the number of possible alignment arrangements is enormous: 6768! / 4790! Without any limitation of search, the number of concept pairs that each alignment heuristic has to consider is over 32.4 million. The alignment suggestion cycle was repeated 5 times. Alignment suggestions were computed once only with the NAME and DEFINITION match heuristics, at the beginning of the sequence, since these suggestions never change (the names and definitions remain the same). However, since every new alignment creates a new bridge across the ontologies, the TAXONOMY match heuristic was re-run 5 times. Its new suggestions, subsequently combined with the other heuristics' suggestions, produced the new suggestions in every run. After 5 runs, the combined heuristics extracted 883 suggestions for manual validation (= 2.72% of the total number of pairs, or 13% of the portion of SENSUS under consideration). Of these, 244 (= 27.6%) were found to be correct, 383 (= 43.4%) were incorrect, and 256 (= 30.0%) were nearly correct (no better alignment existed). Coverage is more difficult to measure; it would involve manually searching the entirety of MIKROKOSMOS to see how many concepts the heuristics should have picked out but did not. Informal estimates, based on browsing the MIKROKOSMOS files and seeing if plausible-looking concepts were picked up for suggested alignment, place the coverage performance fairly high, at over 85%.

Despite the preliminary nature of this work (ongoing research focuses on new alignment suggestion heuristics, more sophisticated validation techniques, and useful tools, such as browsers and editors), we found it very useful during the alignment to be able to focus on only about 13% of the SENSUS terms, and of these, to find just over half to be correct or nearly correct.

3.3 Findings of Alignment

Correct alignments occurred when all three heuristics (name, definition, and taxonomy match) combined well:[2]

|S@change of location,move| = the act of changing your location from one place to another
   superconcepts: (|S@MOTION-PROCESS|)
|M@MOTION-EVENT| = a physical-event in which an agent changing location moves from one place to another
   superconcepts: (|M@CHANGE-LOCATION|)
 	
(COMBINATION SCORE = 4.59, NAME = 26, DEF = 4.50, TAX = 0.20)

|S@foodstuff<food| = a substance that can be used or prepared for use as food
   superconcepts: (|S@food|)
|M@FOODSTUFF| = a substance that can be used or prepared for use as food
   superconcepts: (|M@FOOD| |M@MATERIAL|)
 	
(COMBINATION SCORE = 13.35, NAME = 91, DEF = 10.00, TAX = 0.14)

Sometimes, correct alignments caused errors to be uncovered: Is an archipelago land or sea?

|S@archipelago| = many scattered islands in a large body of water
   superconcepts: (|S@dry land|)
|M@ARCHIPELAGO| = a sea with many islands
   superconcepts: (|M@SEA|)
 	
(COMBINATION SCORE = 1.522, NAME = 131, DEF = 1.33, TAX = 0.00)

However, when studying differences in reasoning between domain modelers and ontologizers, completely equivalent (i.e., correctly aligned) concepts and completely unrelated concepts are not really as interesting as the alignments that are nearly correct. Why do differences exist? How would ontology builders defend (or change) their decisions? How can the reasoning they went through be systematized and used later to guide ontology creation?

Very often, and interestingly, it was quite difficult to decide whether an alignment suggestion was in fact correct, nearly correct, or wrong:

|S@library>bibliotheca| = a collection of literary documents or records kept for reference
  superconcepts: (|S@aggregation|)
|M@LIBRARY| = a place in which literary and artistic materials such as books periodicals newspapers pamphlets and prints are kept for reading or reference an institution or foundation maintaining such a collection
  superconcepts: (|M@ACADEMIC-BUILDING|)

(COMBINATION SCORE = 2.74, NAME = 59, DEF = 3.57, TAX = 0.00)

|S@geisha| = a Japanese woman trained to entertain men with conversation and singing and dancing
  superconcepts: (|S@adult female| |S@Japanese<Asian|)
|M@GEISHA| = a Japanese girl trained as an entertainer to serve as a hired entertainer to men
  superconcepts: (|M@ENTERTAINMENT-ROLE|)

(COMBINATION SCORE = 1.54, NAME = 46, DEF = 2.27, TAX = 0.00)

|S@man<soul| = the generic use of the word to refer to any human being: "It was every man for himself"
  superconcepts: (|S@PERSON|)
|M@HUMAN| = homo sapiens
  superconcepts: (|M@PRIMATE|)

(COMBINATION SCORE = 1.43, NAME = 19, DEF = 0.00, TAX = 0.33)

All three |S@library>bibliotheca|, |S@geisha|, and |S@man<soul| seem to align well to their partners, until one compares their respective superconcepts in SENSUS and MIKROKOSMOS, which focus on different aspects of their nature. A library is *both* a collection of books and a building that houses such a collection; the word can be used to mean both quite naturally, even in the same sentence. A geisha is an adult female with the role of entertainer, a relationship the MIKROKOSMOS builders considered adequately modeled by the use of *is-a*. A human is certainly both a person and a primate, but |S@PERSON| had no equivalent within MIKROKOSMOS (other than, perhaps, |M@HUMAN|, again).

The result of such subsumptive ambiguities has spurred some recent work on what can be called 'concept sense differentiation'. Both in the area of philosophy of knowledge representation (Guarino, 1997, 1998) and of computational lexicography (Pustejovsky, 1995) attempts have been made to distinguish the major kinds of interpretations one can bring to bear on objects. Guarino identifies seven perspectives and Pustejovsky four so-called qualia—for example, one can view an object as a lump of material, as a spatial structure, by its function (if it has one), as a living entity (if it is such), as a social organism (if it is such), and so on. To Guarino, a set of so-called identity criteria can be used to distinguish which sense(s) are relevant at any time: When you smash a glass, is the result still the same glass, for current purposes? Materially, it is; but structurally and functionally, it is obviously not. When after 200 years an organization consists of an entirely new membership, is it still the same organization? Socially, it is; but according to its 'material', its parts, it is not. Taking this approach, |S@geisha| is physically/structurally an |S@adult female| *and is also* functionally an |M@ENTERTAINMENT-ROLE|, for example. (One should not, of course, take this approach too far; although as Heraclitus observed you never step into the same stream twice, one needs to determine carefully which criteria to adopt!)

By carefully comparing the relationships of each aligned or nearly aligned concept with its neighbors in both ontologies, differences of ontological interpretation can be found. Sometimes, as shown above, they are mere errors. Other times, they point to interesting questions of semantic modeling (Visser, Jones, Bench-Capon, & Shave, 1998). Then it becomes appropriate to ask the two ontology builders why such a difference might exist, and how they might motivate or justify their modeling decisions. In particular, such questions may prompt the development of new 'concept senses' and identity criteria.

4. CONCLUSION

As more, and more elaborate, ontologies of all kinds—large-scale cross-domain concept ontologies and detailed and rich domain models—become available, and as we start using, sharing, and comparing ontologies, the need for characterizing ontologies and for establishing some methodology, to the extent possible, becomes increasingly important. Only when we have done so will ontology engineering evolve from an art into a science.

Endnotes

1. Other alignment heuristics have been investigated between ontologies (Agirre, Arregi, Artola, Diaz de Ilarazza, & Sarasola, 1994; Ageno et al., 1994; Rigau & Agirre, 1995; Agangemi, Pisanelli, & Steve, 1998) and between dictionaries and ontologies (Knight & Luk, 1994; Okumura & Hovy, 1994).

2. SENSUS concepts are prefixed with 'S@' and named within upright bars; MIKROKOSMOS concepts are prefixed with 'M@'. The three heuristics' scores ranged as follows: $0 \leq NAME < 150$; $0 \leq DEF < 15$; $0 \leq TAX < 1$.

Acknowledgments

Many thanks to Bruce Jakeway for a lot of work in Section 3, and to comments from Michael Gregory, Graeme Hirst, Fritz Lehmann, Dagobert Soergel, John Sowa, and Mike Uschold for comments on Section 2, and to Steffen Staab for general comments.

References

Agangemi, A., Pisanelli, D., & Steve, G. (1998). Ontology alignment: Experiences with medical terminologies. *Formal Ontology in Information Systems: Proceedings of the First International Conference* (FOIS'98), 61-69.

Ageno, A., Castellon, I., Ribas, F., Rigau, G., Rodriguez, H., & Samiotou, A. (1994). TGE: Tlink generation environment. *Proceedings of the 15th International Conference on Computational Linguistics*, 12-17.

Agirre, E., Arregi, X., Artola, X., Diaz de Ilarazza, A., & Sarasola, K. (1994). Conceptual distance and automatic spelling correction. *Proceedings of the Workshop on Computational Linguistics for Speech and Handwriting Recognition*, 72-80.

Aguado, G., Bañón, A., Bateman, J., Bernardos, S., Fernández, M., Gómez-Pérez, A., Nieto, E., Olalla, A., Plaza, R., & Sánchez, A. (1998). ONTOGENERATION: Reusing domain and linguistic ontologies for Spanish text generation. *Proceedings of the ECAI Workshop on Applications of Ontologies and Problem Solving Methods*, 1-10.

Barwise, J. & Perry. J. (1986). *Situations and Attitudes*. Cambridge, MA: MIT Press.

Bateman, J. A., Kasper, R. T., Moore, J. D., & Whitney, R. A. (1989). A general organization of knowledge for natural language processing: The Penman Upper Model. Unpublished manuscript, Information Sciences Institute, University of Southern California, Marina del Rey, CA. A version of this paper appeared in 1990 as: Upper modeling: A level of semantics for natural language processing. *Proceedings of the 5th International Workshop on Language Generation*.

Borgida, A., Brachman, R. J., McGuinness, D. L., & Alperin Resnick, L. (1989). CLASSIC: A structural data model for objects. *Proceedings of the 1989 ACM SIGMOD International Conference on Management of Data*, 58-67.

Brachman, R. J. (1978). *A Structural Paradigm for Representing Knowledge* (BBN Report No. 3605). Cambridge, MA: Bolt, Beranek, and Newman.

Brachman, R. J., McGuinness, D. L., Patel-Schneider, P. F., & Borgida, A. (1999). "Reducing" CLASSIC to practice: Knowledge representation theory meets reality. *Artificial Intelligence*, 114, 203-237.

Chalupsky, H. (1996). Report on ontology alignment validity test. Unpublished manuscript, Information Sciences Institute, University of Southern California, Marina del Rey, CA.

Davidson, D. (1967). Truth and Meaning. *Synthese*, 17, 304–323.

Dorr, B. J. (1994). Machine translation divergences: A formal description and proposed solution. *Computational Linguistics*, 20, 597–634.

Farquhar, A., Fikes, R., & Rice, J. (1997). The Ontolingua server: A tool for collaborative ontology construction. *International Journal of Human-Computer Studies*, 46, 707–727.

Fellbaum, C. (Ed.). (1998). *WordNet: An Electronic Lexical Database.* Cambridge, MA: MIT Press.

Guarino, N. (1997, March). Some organizing principles for a unified top-level ontology (rev. version). Paper presented at AAAI Spring Symposium on Ontological Engineering, Stanford University, Palo Alto, CA.

Guarino, N. (1998). Some ontological principles for designing upper level lexical resources. *Proceedings of the First International Conference on Language Resources and Evaluation* (LREC-98), 527-534.

Hobbs, J. R. (1985). Ontological promiscuity. *23rd Annual Meeting of the Association for Computational Linguistics: Proceedings of the Conference,* 61-69.

Hovy, E. H. (1996, September). Semi-automated alignment of top regions of SENSUS and CYC. Paper presented to ANSI Ad Hoc Committee on Ontology Standardization, Stanford University, Palo Alto, CA.

Hovy, E. H. (1998). Combining and standardizing large-scale, practical ontologies for machine translation and other uses. *Proceedings of the 1st International Conference on Language Resources and Evaluation* (LREC-98), 535-542.

Hovy, E. H. & Lin, C.Y. (1999). Automated text summarization in SUMMARIST. In M. Maybury & I. Mani (Eds.), *Advances in Automatic Text Summarization,* 81-98. Cambridge, MA: MIT Press.

Iwasaki, Y., Fraquhar, A., Fikes, R., & Rice, J. (1997). *A Web-Based Compositional Modeling System for Sharing of Physical Knowledge.* Nagoya: Morgan Kaufmann.

Jasper, R., & Uschold, M. (1999). A framework for understanding and classifying ontology applications. *Proceedings of the Twelfth Workshop on Knowledge Acquisition, Modeling and Management.* Available: <http://sern.ucalgary.ca/ KSI/KAW/KAW99/papers/Uschold2/final-ont-apn-fmk.pdf> [2001, October 9].

Knight, K., & Luk, S.K. (1994). Building a large-scale knowledge base for machine translation. *Proceedings of the Twelfth National Conference on Artificial Intelligence,* 773-778.

Knight, K., Chander, I., Haines, M., Hatzivassiloglou, V., Hovy, E. H., Iida, M., Luk, S. K., Whitney, R.A., & Yamada, K. (1995). Filling knowledge gaps in a broad-coverage MT system. *Proceedings of the 14th International Joint Conference on Artificial Intelligence,* 1390-1396.

Lehmann, F. (1997, March). Reworked alignment of top regions of CYC and SENSUS. Paper presented to ANSI Ad Hoc Committee on Ontology Standardization, Stanford University, Palo Alto, CA.

Lenat, D. B. (1995). CYC: A large-scale investment in knowledge infrastructure. *Communications of the ACM,* 38(11), 32–38.

MacGregor, R. (1991). Inside the LOOM descriptive classifier. *SIGART Bulletin,* 2(3), 70–76.

Mahesh, K. (1996). *Ontology Development for Machine Translation: Ideology and Methodology* (CRL Report MCCS-96-292). Las Cruces: New Mexico State University.

McDermott, D. V. (1981). Artificial intelligence meets natural stupidity. In J. Haugland (Ed.), *Mind Design.* Cambridge: MIT Press.

McGuinness, D. L., Fikes, R., Rice, J., & Wilder, S. (2000). An environment for merging and testing large ontologies. *Proceedings of the Seventh International Conference on Principles of Knowledge Representation and Reasoning.*

Montague, R. (1974). *Formal Philosophy.* New Haven, CT: Yale University Press.

Nirenburg, S., Raskin V., & Onyshkevych, B. (1995). Apologiae ontologia. *Proceedings of the International Conference on Theoretical and Methodological Issues.*

Noy, N. F., & Hafner, C. D. (1997). The state of the art in ontology design: A survey and comparative review. *AI Magazine,* 18(3), 53–74.

Noy, N. F., & Musen, M. A. (1999a). SMART: Automated support for ontology merging and alignment. *Proceedings of the Twelfth Workshop on Knowledge Acquisition, Modeling and Management.* Available: <http://sern.ucalgary.ca/ KSI/KAW/KAW99/papers/Fridman1/NoyMusen.pdf [2001, October 9]. Also available: <http://smi-web.stanford.edu/pubs/SMI_Abstracts/SMI-1999-0813.html [2001, October 9].

Noy, N. F., & Musen, M. A. (1999b). An algorithm for merging and aligning ontologies: Automation and tool support. *Proceedings of the Workshop on Ontology Management at the Sixteenth National Conference on Artificial Intelligence,* 1201-1206. Also available: <http://www-smi.stanford.edu/pubs/SMI_Reports/SMI-1999-0799.pdf> [2001, October 9].

Okumura, A., & Hovy, E. H. (1994). Ontology concept association using a bilingual dictionary. *Proceedings of the 1st Conference of the Association for Machine Translation in the Americas,* 177-184.

Pustejovsky, J. (1995). *The Generative Lexicon.* Cambridge, MA: MIT Press.

Rigau, G., & Agirre, E. (1995). Disambiguating bilingual nominal entries against WordNet. *Proceedings of the 7th ESSLLI Symposium,* 16-23.

Schank, R. C., & Abelson, R. P. (1977). *Scripts, Plans, Goals, and Understanding.* Hillsdale, NJ: Erlbaum.

Sowa, J. F. (2000). *Knowledge Representation: Logical, Philosophical, and Computational Foundations.* Pacific Grove, CA: Brooks/Cole.

Swartout, W. R., Patil, R., Knight, K., & Russ, T. (1996). Toward distributed use of large-scale ontologies. *Proceedings of Tenth Knowledge Acquisition for Knowledge-based Systems Workshop.* Available: <http://ksi.cpsc.ucalgary.ca/ KAW/KAW96/swartout/Banff_96_final_2.html> [2001, October 9].

Uschold, M., Clark, P., Healy, M., Williamson, K., & Woods, S. (1998). Ontology reuse and application. *Formal Ontology in Information Systems: Proceedings of the First International Conference* (FOIS'98), 179-192.

Van Zyl, J., & Corbett, D. (2000). A framework for comparing the use of a linguistic ontology in an application. *Proceedings of the ECAI Workshop on Applications of Ontologies and Problem Solving Methods,* 35-42.

Visser, P., Jones, D., Bench-Capon, T., & Shave, M. (1998). Assessing heterogeneity by classifying ontology mismatches. *Formal Ontology in Information Systems: Proceedings of the First International Conference* (FOIS'98), 148-162.

Wille, R. (1997). Conceptual graphs and formal concept analysis. *Conceptual Structures: Fulfilling Peirce's Dream: Fifth International Conference on Conceptual Structures,* 290–303.

Chapter 7

Identity and Subsumption

Nicola Guarino
Institute for Systems Science and Biomedical Engineering of the Italian National Research Council (LADSEB-CNR), Padova, Italy

Christopher Welty
Computer Science Department, Vassar College, Poughkeepsie, NY, USA

Abstract:
    The intuitive simplicity of the so-called *is-a* (or *subsumption*) relationship has led to widespread ontological misuse. Where previous work has focused largely on the semantics of the relationship itself, we concentrate here on the ontological nature of its *arguments*, in order to tell whether a single *is-a* link is ontologically well-founded. For this purpose, we introduce some techniques based on the philosophical notions of *identity, unity,* and *essence*, which have been adapted to the needs of taxonomy design. We demonstrate the effectiveness of these techniques by taking real examples of poorly structured taxonomies and revealing cases of invalid generalization.

1. INTRODUCTION

Taxonomies based on a partial-ordering relation commonly known as *is-a*, class inclusion, or subsumption have become an important conceptual modeling tool for database schemas, knowledge-based systems, and semantic lexicons. Properly structured taxonomies help bring substantial order to conceptual models, are particularly useful in presenting limited views for human interpretation, and play a critical role in reuse and integration tasks. Improperly structured taxonomies have the opposite effect, making models confusing and difficult to reuse or integrate.

Many previous efforts at providing some clarity in organizing taxonomies have focused on the semantics of the subsumption relationship (Brachman, 1983), on various kinds of related relations (generalization, specialization, subset hierarchy) (Storey, 1993), or on its role in the more general framework of data abstractions (Goldstein & Storey, 1999). Our approach differs in that we focus on the arguments (i.e., the properties) involved in the subsumption relationship, rather than on the semantics of the relationship itself. The latter is taken for granted, as we take the statement "ψ subsumes φ" to mean that, *necessarily:*

(1) $\forall x \; \varphi(x) \rightarrow \psi(x)$

The modal reading of the above formula is an important qualification: We take subsumption as an *ontological constraint*; therefore we assume that it must hold in all possible worlds. Indeed, we believe that modal necessity is what distinguishes—within a particular conceptualization—an ontological truth from a contingent assertion. So we focus here on *necessary subsumption*. Our task will be to verify its plausibility on the basis of the *ontological nature* of its arguments.

This chapter is organized as follows. First, we clarify some major issues lying behind the generic notion of identity, and the related notions of unity and essence. Then we show how these notions impose ontological constraints on the subsumption relationship, and discuss some concrete examples of problems and misconceptions concerning subsumption taxonomies.

Most of the ideas discussed here have been introduced in previous papers (Guarino & Welty 2000a, 2000b, in press). The present work does however refine and simplify most of the core definitions presented in the past, and aims at offering a self-contained overview for what concerns the ontological analysis of the subsumption relationship.

2. IDENTITY, UNITY, AND ESSENCE

Identity is one of the most fundamental notions in ontology, yet the related issues are very subtle, and isolating the most relevant ones is not an easy task; see (Hirsch, 1982) for an account of the identity problems of ordinary objects, and (Noonan, 1993) for a collection of philosophical papers in this area. In particular, the relationship between *identity* and *unity* appears to be crucial for our interest in ontological analysis. These notions are different, albeit closely related and often confused under a generic notion of identity. Strictly speaking, identity is related to the problem of distinguishing a specific instance of a certain class from other instances by means of a *characteristic property*, which is unique for *it* (that *whole* instance). Unity, on the other hand, is related to the problem of distinguishing the *parts* of an instance from the rest of the world by means of a *unifying relation* that binds them together (not involving anything else).

For example, asking "Is that my dog?" would be a problem of identity, whereas asking "is the collar part of my dog?" would be a problem of unity. As we shall see, the two notions are complementary: When something can be both recognized as a whole and kept distinct from other wholes then we say that it is an *individual* and can be counted as *one*.

The actual conditions we use to support our answers to these questions for a certain class of things vary from case to case, depending on the properties holding for these things. If we find a condition that consistently supports identity or unity judgments for *all* instances of a certain property, then we say that such property *carries* an identity or a unity condition.

These notions encounter problems when time is involved. The classical one is that of *identity through change*: In order to account for common sense (i.e., for the way we, as cognitive agents, interact with the world around us), we need to admit that an individual may keep its identity while exhibiting different properties at different times. But which properties can change, and which must not? And how can we re-identify an instance of a certain property after some time? The former issue leads to the notion of an *essential property*, on which we base the definition of *rigidity*; the latter is related to the distinction between *synchronic* and *diachronic* identity. Both issues will be discussed below.

Before going on, it is important to make clear that all the assumptions related to the notions above depend on our *conceptualization* of the world (Guarino, 1998). For example, the decision as to whether cats keep their identity after losing their tail, or whether statues are identical with the marble they are constituted of, are ultimately the result of our sensory system, our culture, and so on. The examples we shall use in this chapter concerning the ontological nature of certain properties (e.g., *STUDENT*, *SPHERICAL*) are merely indicative of our own intuitions. The aim of the present analysis is not so much to discuss these assumptions, but rather to clarify the formal tools needed to make them explicit, and to explore their logical consequences. These formal tools form the core of a methodology for *ontology-driven conceptual modeling* called OntoClean, which is discussed in more detail elsewhere (Guarino & Welty, in press).

3. THE FORMAL FRAMEWORK

In this section we present a formal analysis of the basic notions discussed above, and we introduce a set of *meta-properties* that represent the behavior of a property with respect to these notions. Our goal is to show how these meta-properties impose relevant ontological constraints on the subsumption relationship.

In the following, we shall denote meta-properties by bold letters preceded by the sign "+", "-" or "~", whose meaning will be described for each meta-property. We use the notation φ^M to indicate that the property φ has the meta-property **M**.

We shall adopt a simple temporal logic, where all predicates are temporally indexed by means of an extra argument. If the time argument is omitted for a certain predicate P, then the predicate is assumed to be time invariant, that is $\exists t P(x,t) \rightarrow \forall t P(x,t)$. Our domain of quantification will be that of *possibilia*: This means that we include all possible entities, independent of their actual existence (Lewis, 1983). Therefore, we shall quantify over a constant domain in every possible world. Worlds will be considered "histories" (temporally ordered sequences of maximal states of affairs) rather than "snapshots," and we shall consider all of them as equally accessible. As a result, we shall adopt the simplest quantified modal logic, namely S5 plus the Barcan Formula (Hughes & Cresswell, 1996).

For example, the property *UNICORN* will not be empty in our world, although no instance has actual existence there. Actual existence is therefore different from existential quantification ("logical existence"), and will be represented by the temporally indexed predicate E(x,t), meaning that x has actual existence at time t (Hirst, 1991).

Finally, in order to avoid trivial cases in our meta-property definitions, we shall implicitly assume the property variables as restricted to *discriminating properties*, properties φ such that $\Diamond \exists x\, \varphi(x) \land \Diamond \exists x\, \neg\varphi(x)$. In other words, discriminating properties are properties for which there is possibly something which exhibits that property, and possibly something that does not exhibit that property; they are neither tautological nor vacuous.

3.1 Essential Properties and Rigidity

Before addressing the core issues of identity and unity, it is useful to clarify the notion of *essential property*, and to introduce a set of related meta-properties.

Definition 1 A property φ holding for a certain individual a in a certain state of affairs at time t is said to be *essential* to a iff it necessarily holds for a at every possible time in every possible world, i.e.,

(2) $\square \, \forall t \, \varphi(a, t)$

Examples of essential properties for a human being would be PERSON and HAVING A BRAIN.

Definition 2 A property φ is *rigid* iff, necessarily, it is essential to *all* its instances, i.e.,

(3) $\square \, \forall x t \, (\varphi(x, t) \rightarrow \square \, \forall t \, '\varphi(x, t'))$

For example, HAVING A BRAIN would be essential to human beings but not to all its instances, since it is not essential to, say, a dead corpse. On the other hand, PERSON can be safely be taken as rigid. Other examples of non-rigid properties could be TIRED or STUDENT.

Definition 3 A property φ is *non-rigid* iff it is not rigid, that is, there is at least one instance such that φ is not essential to it:

(4) $\lozenge \, \exists x \, (\varphi(x, t) \wedge \lozenge \, \exists t \, '\neg\varphi(x, t'))$

Non-rigidity can be further restricted as follows:

Definition 4 A property φ is *anti-rigid* iff all its instances are such that φ is not essential to them:

(5) $\square \, (\forall x t \, \varphi(x, t) \rightarrow \lozenge \, \exists t \, '\neg\varphi(x, t'))$

Consider for example the properties SPHERICAL and STUDENT. The former is non-rigid but not anti-rigid, since it may be the case that something (like a lump of clay) is spherical by accident, but it is also possible that there are things that are essentially spherical (spheres, for instance). The STUDENT property, on the other hand, appears to be anti-rigid, since it is always possible for a student to become a non-student while being the same individual.

Rigid properties are marked with the meta-property +**R**, non-rigid properties are marked with -**R**, and anti-rigid properties with ~**R**. Note that rigidity as a meta-property is not "inherited" by sub-properties of properties that carry it, so the markings PERSON[+R] and STUDENT[~R] are perfectly consistent with the fact that PERSON subsumes STUDENT.

3.2 Identity and Identity Conditions

Before discussing the formal structure of identity conditions (ICs), some clarifications about their intuitive meaning may be useful. If we say, "Two persons are the same if they have the same SSN," we seem to create a puzzle: How can they be *two* if they are the same? The puzzle can be solved by recognizing that two (incomplete) descriptions of a person (like two records in different databases) can be different while referring to the same individual. The statement "two persons are the same" can be therefore rephrased as "two descriptions of a person refer to the same object". A description can be seen as a set of properties that apply to a certain object. Our intuition is that two incomplete descriptions denote the same object if they have an identifying property in common.

Depending on whether the two descriptions hold at the same time, we distinguish between *synchronic* and *diachronic* ICs. The former are needed to tell, for example, whether the statue is identical with the marble it is made of, or whether a hole is identical with its filler (Casati & Varzi, 1994), while the latter allow us to re-identify things over time.

In the philosophical literature, an *identity criterion* is generally defined as a condition that is both necessary and sufficient for identity. According to (Strawson, 1959), a property φ *carries* an IC iff the following formula holds for a suitable ρ:

(6) $\quad \varphi(x) \wedge \varphi(y) \rightarrow (\rho(x,y) \leftrightarrow x=y)$

Since identity is an equivalence relation, it follows that ρ restricted to φ must also be an equivalence relation. For example, the property PERSON can be seen as carrying an IC if relations like *having-the-same-brain* or *having-the-same-SSN* are assumed to satisfy (6).

Properties carrying an IC are called *sortals* (Strawson, 1959). In many cases, their linguistic counterparts are *nouns* (e.g., Apple), while non-sortals correspond to *adjectives* (e.g., Red). Distinguishing sortals from non-sortals is of high practical relevance for conceptual modeling, as we tend to naturally organize knowledge around nouns.

When trying to use (6) for conceptual modeling purposes we encountered however a number of problems. First, the nature of the ρ relation remains mysterious: What makes it an IC, and how can we index it with respect to time to account for the difference between synchronic and diachronic identity? Second, it only accounts for identity *under a certain property*: In principle it may be that $x=y$ without being $\varphi(x,t) \wedge \varphi(y,t')$ (with $t \neq t'$), but the formula (6) does not help us in this case. Third, deciding whether a property carries an IC or not may be difficult, since finding a ρ that is both necessary *and* sufficient for identity is often hard, especially for natural kinds and artifacts.

3.3 The "Sameness" Relation

Our intuition is that the nature of the ρ relation in (6) is based on the "sameness" of a certain *property*, which is unique to a specific instance. Suppose we stipulate, for example, that two persons are the same iff they have the same brain: The reason why this *relation* can be used as an IC for persons lies in the fact that a property like "having this particular brain" is an *identifying property*, since it holds exactly for one person.

Identifying properties can be seen as relational properties, involving a *characteristic relation* between a class of individuals and their *identifying characteristics*.[1] In the above example, brains are taken as identifying characteristics of persons. Such characteristics can be internal to individuals themselves (parts or qualities) or external to them (other "reference" entities). So two things can be the same because they have some parts or qualities in common, or because they are related in the same way to something else (for instance, we may say that two material objects are the same if they occupy the same spatial region). Of course, an individual's characteristic cannot be identical to the individual itself, so a characteristic relation must be *irreflexive*.

This means that, if χ denotes a suitable characteristic relation for φ, we can assume:

(7) $\rho(x,y)=\forall z(\chi(x,z) \leftrightarrow \chi(y,z))$

The scheme (6) becomes therefore:

(8) $\varphi(x) \wedge \varphi(y) \rightarrow (\forall z(\chi(x,z) \leftrightarrow \chi(y,z)) \leftrightarrow x=y)$

For instance, if we take φ as the property of being a set, and χ as the relation "has-member", this scheme tells us that two sets are identical iff they have the same members.

An important advantage of (8) over (6) is that it is based on a characteristic relation χ holding separately for x and y, rather than on a relation ρ holding between them. This allows us to take time into account more easily, clarifying the distinction between *synchronic* and *diachronic* identity:

(9) $E(x,t) \wedge E(y,t') \wedge \varphi(x,t) \wedge \varphi(y,t') \rightarrow (\forall z(\chi(x,z,t) \leftrightarrow \chi(y,z,t')) \leftrightarrow x=y)$

We shall have a synchronic criterion if $t=t'$, and a diachronic criterion otherwise. Note that accounting for the difference between synchronic and diachronic identity would be difficult with (6): We may think of adding two temporal arguments to the ρ relation, but in this case its semantics would become quite unclear, being a relation that binds together two entities at different times. Note also that synchronic identity criteria are weaker than diachronic ones. For instance, the sameness of spatial location is usually adopted as a synchronic identity criterion for material objects, but of course it does not work as a diachronic criterion.

A possible criticism of (9) is that it looks circular, since it defines the identity between x and y in terms of the identity between something else (in this case, the identifying characteristics z common to x and y). However, as observed by Lowe (1998, p. 45), we must keep in mind that *ICs are not definitions*, as identity is a primitive. This means that the circularity of identity criteria with respect to the very notion of identity is just a fact of life: Identity can't be defined. Rather, we may ask ICs to be *informative*, in the sense that identity conditions must be non-circular with respect to the properties involved in their definition. For instance, Lowe points out that Davidson's identity criterion for events, stating that two events are the same if they have the same causes and they are originated by the same causes (Davidson, 1980), is circular in this sense since it presupposes the identity of causes, which are themselves events. In many cases, however, even this requirement cannot easily be met, and we must regard ICs as simple constraints.

For brevity, the formula (9) above can be rewritten as:

(10) $E(x,t) \wedge E(y,t') \wedge \varphi(x,t) \wedge \varphi(y,t') \rightarrow (\Sigma\chi(x,y,t,t') \leftrightarrow x=y)$

where $\Sigma\chi$ is a *sameness formula*, defined as

(11) $\Sigma\chi(x,y,t,t')=_{def}\forall z(\chi(x,z,t) \leftrightarrow \chi(y,z,t'))$

We may conclude therefore that an IC for φ is a sameness formula $\Sigma\chi$ that satisfies (10) and is based on a suitable characteristic relation χ. It is safe however to make sure that the IC really depends on φ, imposing a *non-triviality constraint* such as:

(12) $\neg\forall xy(\Sigma\chi(x,y,t,t') \leftrightarrow x=y)$

For instance, suppose that χ is the *proper-part* relation: In this case the statement $\forall xy(\Sigma\chi(x,y,t,t') \leftrightarrow \text{x=y})$ would represent the extensionality principle, which says that two (non-atomic) entities are the same iff they have the same proper parts. Without the constraint (12), any property holding for non-atomic entities would trivially carry an identity criterion if the extensionality principle was assumed.[2]

3.4 Local and Global Identity Conditions

The formula (10) above may hold for *rigid* properties like PERSON, as well as for *non-rigid* properties like STUDENT or TIRED. In the latter case, $\Sigma\chi$ may act only as *local* IC for φ, as we can't be sure it also accounts for identity among entities that are not both instances of φ. Consider, for example, the properties CATERPILLAR and BUTTERFLY: In this case, a formula $\Sigma\chi$ based (for instance) on the sameness of a certain wing pattern could count as an IC for butterflies, but it would not account for the identity relation holding, at different times, between butterflies and caterpillars.

It seems useful therefore to distinguish between *local* and *global* ICs. The following formal definitions are intended to account for such a distinction, as well as for the related issues concerning the way ICs are inherited along subsumption hierarchies. They refine some previous definitions reported in Guarino & Welty (2000a, 2000b, in press).

Definition 5 Let φ be a property, and $\chi(x,z,t)$ a non-trivial characteristic relation satisfying (12), and such that $\square(\varphi(x,t) \rightarrow \exists z\chi(x,z,t))$. Then

- φ *carries a local identity condition* $\Sigma\chi$ iff, necessarily:

    (13) $E(x,t) \wedge E(y,t') \wedge \varphi(x,t) \wedge \varphi(y,t') \rightarrow (\Sigma\chi(x,y,t,t') \leftrightarrow x=y)$

- φ *carries a global identity condition* $\Sigma\chi$ iff, necessarily:

    (14) $E(x,t) \wedge E(y,t') \wedge \varphi(x,t) \rightarrow (\Sigma\chi(x,y,t,t') \leftrightarrow x=y)$

Since ICs can be inherited along subsumption hierarchies, it is useful to distinguish between *supplying* an IC and just *carrying* it:

Definition 6 A property φ is a *sortal* iff it carries a (local or global) IC. Sortals are marked with the meta-property **+I**.

Definition 7 A property $\varphi \in \mathbf{O}$ *supplies* a (local or global) IC $\Sigma\chi$ in $\mathbf{O}$ iff (i) $\mathbf{O}$ is the set of *explicit properties* introduced in a certain ontology (i.e., corresponding to predicate names); (ii) φ carries $\Sigma\chi$; and (iii) $\Sigma\chi$ is not carried by all the properties in $\mathbf{O}$ directly subsuming φ. This means that, if φ inherits different (but compatible) ICs from multiple properties, it still counts as supplying an IC (see section 3.6).

Properties supplying global identity are marked with the meta-property **+G**;[3] those supplying local identity are marked with **+L**.

Definition 8 A property φ is a *type* iff it supplies a global IC.

Let us now introduce an important principle, adapted from [Lowe 89]:

Sortal Expandability Principle (SEP). If two entities are identical, they must be instances of a common sortal that *accounts* for their identity, i.e., it carries an IC they satisfy. In a *well-founded ontology*, there must be a *unique* sortal that *supplies* such IC.

On the basis of this principle, we can prove the following:

Theorem 1 In a well-founded ontology, *types are rigid properties.*

To see this, let's suppose that a type φ, supplying a certain IC $\Sigma\chi$, is not rigid. This means that, for some x, y, and t, with $x=y$, there are two worlds w_1 and w_2 such that $\varphi(x,t)$ holds in w_1 and $\neg\varphi(y,t')$ holds in w_2. Because of the sortal expandability principle, there must be a unique sortal ψ that holds in the two situations, and that supplies $\Sigma\chi$. But, by hypothesis, $\Sigma\chi$ is supplied by φ. Because of the uniqueness constraint, it must be $\psi=\varphi$. This means that φ must be rigid, contradicting the original hypothesis.

Recognizing types is of utmost importance in ontology design. Since they must be rigid in order to supply global IC, they represent *invariant properties* that characterize the nature of a domain element by supplying identity criteria to it. If we assume Quine's motto "No entity without identity," this implies that every element of our domain of discourse must be an instance of a type. Types and other property kinds defined on the basis of the meta-properties discussed here have been presented in more detail (although with a few formal differences) in Guarino & Welty (2000a).

3.5 Heuristics for Identity

Unfortunately, recognizing that a property carries a *specific* IC is often difficult in practice. However, in many cases it suffices to recognize whether a property carries *some* IC (being therefore a sortal) or not, without telling exactly *which* IC. In these cases, we may want to check for some *minimal* ICs, which are (only)-necessary or (only)-sufficient for identity, but close enough to "true" ICs. If none of these weak conditions holds for a given property, we may safely conclude it is not a sortal. Otherwise, we may have some heuristic evidence that some "true" IC exists.

Only-necessary and only-sufficient conditions for global identity can be defined by considering separately the two senses of the double implication in (13). We need however to assume in advance that φ is a rigid property, and to make sure to exclude trivial cases (in the following, (16) is needed to guarantee that the second literal in (15) is relevant and not tautological).

Definition 9 A *necessary global identity condition* for a rigid property φ is a formula $\Sigma\chi$, satisfying (11) and (12), such that:

(15) $\varphi(x,t) \wedge x=y \rightarrow \Sigma\chi(x,y,t,t')$

(16) $\neg\forall xy(\varphi(x,t) \wedge \varphi(y,t') \rightarrow \Sigma\chi(x,y,t,t'))$

Definition 10 A *sufficient identity condition* for a rigid property φ is a formula Σχ, satisfying (11) and (12), such that:

(17) $\varphi(x,t) \wedge \Sigma\chi(x,y,t,t') \rightarrow x=y$

Besides being useful for recognizing sortals, minimal ICs have also a practical relevance in taxonomy design, since of course they also follow the Identity Disjointness Constraint below. As we shall see, this means that in practice we can assume a property carries identity on the basis of the evidence given by the minimal ICs and use them in place of true ICs to constrain the taxonomy.

3.6 Inheriting Multiple Identity Conditions

As described above, ICs can be inherited along subsumption hierarchies. They can also *specialize* along hierarchies, in the sense that new identity criteria can be supplied by a given property in addition to those inherited from the subsuming properties. Consider for instance the domain of abstract geometrical figures, where the property *POLYGON* subsumes *TRIANGLE*. A necessary and sufficient IC for polygons is "having the same edges and the same angles." On the other hand, an *additional* necessary and sufficient IC for triangles is "having two edges and their internal angle in common" (note that this condition is only-necessary for polygons). Again, the property *EQUILATERAL TRIANGLE* may inherit from *REGULAR POLYGON* the IC "having the same edges," while also inheriting the ICs carried by *TRIANGLE*.

As this example shows, nothing prevents sortals from using multiple inheritance to form tangled hierarchies, at least in principle. However, the presence of explicit ICs attached to them imposes an important (as well as natural) constraint on the inheritance of multiple ICs, which we shall call the *Identity Disjointness Constraint* (IDC):

(18)   *Properties carrying incompatible ICs are necessarily disjoint.*

We shall see in the following how this simple principle, whose philosophical implications have been discussed in the seminal work by Lowe (1989, 1998), has a deep impact on apparently innocent taxonomic assumptions.

3.7 Unity

We have discussed and formalized the notion of unity in some detail in previous work (Guarino & Welty, 2000b). At the time the present chapter was being finalized, a new formalization was proposed (Gangemi et al., 2001), which overcomes some difficulties of the previous one.[4] Since the new formalization does not affect the main point of this chapter, which is about the constraints that identity and unity conditions impose on the subsumption relation, we stick here to our former formalization, which seems also easier to grasp.

The notion of unity is closely tied to that of parthood, so that we need to introduce some basic axioms and definitions. We adopt a time-indexed mereological relation $P(x,y,t)$, meaning that x is a (proper or improper) part of y at time t, satisfying the minimal set of axioms and definitions (adapted from Simons, 1987, p. 362) shown in table 1. Differently from Simons, this mereological relation will be taken as completely general, holding on a domain which includes individuals, collections, and amounts of matter.

$PP(x,y,t) =_{def} P(x,y,t) \wedge \neg x=y$	(proper part)
$O(x,y,t) =_{def} \exists z(P(z,x,t) \wedge P(z,y,t))$	(overlap)
$P(x,y,t) \wedge P(y,x,t) \rightarrow x=y$	(antisymmetry)
$P(x,y,t) \wedge P(y,z,t) \rightarrow P(x,z,t)$	(transitivity)
$PP(x,y,t) \rightarrow \exists z(PP(z,y,t) \wedge \neg O(z,x,t))$	(weak supplementation)

Table 1. Axiomatization of the part relation.

Definition 11 At a given time t, an entity x is a *whole under* R if R is an equivalence relation (called *unifying relation*) such that:

(19) $\forall y(P(y,x,t) \rightarrow \forall z(P(z,x,t) \leftrightarrow R(z, y,t)))$
(20) $\neg(PP(y,x,t) \wedge PP(z,x,t) \leftrightarrow R(z, y,t))$

We can read the above definition as follows: *At time t, each part of x must be bound by R to all other parts and to nothing else.* (19) expresses a condition of *maximal self-connectedness* according to a suitable relation of "generalized connection" *R*. (20) is a non-triviality condition on *R*, that avoids considering any mereological sum as a contingent whole.

Depending on the ontological nature of such relation, we may have different kinds of unity. For example, we may distinguish *topological unity* (a piece of coal, a lump of coal), *morphological unity* (a ball, a constellation), *functional unity* (a hammer, a bikini). As a further example, an atomic object (i.e., an object with no proper parts), is a whole under the identity relation. As these examples show, nothing prevents a whole from having parts that are themselves wholes (under different unifying relations). Indeed, a *plural whole* can be defined as a whole which is a mereological sum of wholes.

According to Definition 11, an entity may be a whole only in a particular possible world at a certain time. Consider for instance an isolated piece of clay. This certainly has a certain topological unity, which is however lost as soon as we attach it to a much larger piece: The original piece of clay is not a whole any more, while the new piece is.

A stronger and more useful notion of whole can be introduced by assuming that the same conditions for unity must necessarily hold for an object, i.e., by assuming unity as an *essential* property:

Definition 12 An entity x is an *essential whole under R*, if, necessarily, it is always a whole under *R*.

We are now in the position to state the following:

Definition 13 A property φ *carries a unity condition* (UC), or simply *carries unity*, if there is a common unifying relation *R* such that all its instances are essential wholes under *R*. Properties carrying unity are marked with the meta-property **+U** (**-U** otherwise).

Within properties that do not carry unity, we distinguish properties that do not carry a *common* UC for all their instances from properties all of whose instances are not wholes. An example of the former kind may be LEGAL AGENT, all of whose instances are wholes, although with different UCs (some legal agents may be people, some companies). AMOUNT OF MATTER is usually an example of the latter kind, since none of its instances can be wholes (assuming that a single molecule does not count as an *amount* of matter). Therefore we define:

Definition 14 A property φ carries *anti-unity* (marked with the meta-property ~U) if no instance of it is an essential whole.

4. CONSTRAINTS ON SUBSUMPTION

Let us see now how identity and unity affect the subsumption relationship. Our point is to check the consistency and the ontological plausibility of a subsumption relationship between two properties on the basis of their behavior with respect to identity and unity.

A first important constraint has been introduced in Section 3.6. It follows that a constraint similar to the Identity Disjointness Constraint holds also for Unity, so that we can state:

(21)   *Properties with incompatible ICs/UCs are necessarily disjoint.*

Indeed, the statement above is just the consequence of the fact that having a certain IC or UC is an *essential property*, and incompatible essential properties must be necessarily disjoint. In many cases, just considering the essential properties associated to a given property (independently of any considerations related to identity or unity) is enough to conclude that a certain subsumption link is invalid, just because the two arguments are associated to incompatible essential properties (Akiba, 2000). For instance, to see that a vase is not an amount of matter, we may just consider that the property of having a certain shape is essential for vases, while the same property is "anti-essential" for amounts of matter, in the sense that any amount of matter can possibly have a different shape.

Besides this, we have several constraints involving the meta-properties we have introduced. Let us represent with $\neg\,(\psi \to \varphi)$ a constraint stating that property φ can't subsume ψ. The following constraints are a simple consequence of our definitions:

(22) $\neg\,(\psi^{+R} \to \varphi^{\sim R})$
(23) $\neg\,(\psi^{-I} \to \varphi^{+I})$
(24) $\neg\,(\psi^{-U} \to \varphi^{+U})$
(25) $\neg\,(\psi^{+U} \to \varphi^{\sim U})$

5. SOME PROBLEMATIC SUBSUMPTION RELATIONSHIPS

Let us finally examine some examples of subsumption relationships in the light of the above discussion (table 2). All these examples appear acceptable at a first sight, but immediately become problematic as soon as identity and unity are taken into account.

Note that a complete ontological analysis of these examples would go much beyond the scope of this chapter, so we shall keep the discussion below rather informal. A far more in-depth example is available in Guarino & Welty (2000c).

1	A physical object is an amount of matter	Pangloss
2	An amount of matter is a physical object	WordNet
3	An organization is a group	WordNet
4	An organization is both a social being and a group	CYC
5	A place is a physical object	Mikrokosmos, WordNet
6	A window is both an artifact and a place	Mikrokosmos
7	A person is both a physical object and a living being	Pangloss
8	An animal is both a solid tangible thing and a perceptual agent	CYC
9	A car is both a solid tangible thing and a physical device	CYC
10	A communicative event is a physical, a mental, and a social event	Mikrokosmos

Table 2. Problematic subsumption relationships in some current ontologies.

Examples 1 and 2 clearly represent incompatible ontological commitments, unless we assume that amounts of matter and physical objects collapse into the same concept. As usual, the problem is that the underlying commitment has not been made explicit enough by the authors of these ontologies, and we only have the taxonomy to judge what the intended meaning of the terms "physical object" and "amount of matter" is. The analysis we have presented can help us to solve the puzzle, at least with respect to our own understanding of these terms. According to common sense, amounts of matter can be assumed to carry an extensional IC (two amounts of matter are the same iff they have the same parts) and anti-unity (since every amount of matter is not an essential whole). Physical objects allow for two possible options concerning their identity: In one account, they seem to have a non-extensional IC, since a physical object may keep its identity after replacing or removing some of its parts (e.g., a car with new tires); in a different account, they may have an extensional IC, if we assume that two physical objects are different if they don't have the same parts. In any case, it seems natural to assume that physical objects *do* have unity, since we normally *count* them. In conclusion, the property AMOUNT OF MATTER can be labeled +I, ~U, while PHYSICAL OBJECT can be labeled +I, +U. This means that example 1 violates constraint (25), and example 2 violates constraint (24). Moreover, constraint (21) would be violated in both cases under the assumption of non-extensionality for physical objects. Our conclusion is that physical objects are *constituted* by amounts of matter, but they are *disjoint* from amounts of matter.

Examples 3 and 4 are similar. To analyze them, we have to decide what the IC for group (of people) is. If we admit that a group of people loses its identity when a member is replaced or deleted (as we believe is plausible), then the IC of ORGANIZATION becomes incompatible, since we clearly admit that organizations can change members. In both examples, therefore, constraint (21) is violated. Examining why this is so reveals our assumption that an organization is more than just a group of people: In fact, the same group of people could *constitute* different organizations.

Examples 5 and 6 include the notion of place. We have (at least) two possible ontological choices regarding this notion: (i) We think of a place as a region of space (either absolute or relative space—the issue doesn't matter here), adopting therefore an extensional IC (two regions of space are the same iff they have the same parts); (ii) We think of a place as a *feature* of something else (for instance, a hole in a wall). In the latter case, it seems plausible to give up the assumption of extensionality (if we think that the same hole can change its size) and to introduce an assumption of unity.

Let us now consider example 5. If we take physical objects as non-extensional (as above), then option (i) violates constraint (21). Option (ii) is consistent with (21), but this is a case where a further check of the *essential properties* of places and physical objects would be useful. For instance, if we take physical objects as being (essentially) *material* objects, then we have an obvious inconsistency. So, to account for example 5, we need to allow for *immaterial* physical objects. A further issue concerns, however, the ontological assumptions about *dependence*. If we take physical objects as ordinary objects, like a table or a glass, then we usually assume that they are (essentially) independent of everything else, i.e., they can actually exist, even if nothing else does actually exist. Such an assumption would be incompatible with option (ii), since a feature (like a hole) is an essentially dependent object, in the sense that it cannot exist unless something else (its host) exists. In conclusion, example 5 is consistent only if we consider places as features, and take physical objects in a very general sense, with no commitments regarding their materiality and their independence.

As for example 6, this is a classical case where multiple subsumption risks being improperly used to account for lexical polysemy. If a window is assumed to be a *material* physical object (e.g, a suitably framed glass pane), then it cannot be a place at the same time. So there is a multiple *lexical* link that links WINDOW with its hypernyms, but this can't correspond to a subsumption link. The solution is to introduce two separate nodes, WINDOW-ARTIFACT and WINDOW-PLACE, which account for the two meanings of the word.

Examples 7 and 8 resemble each other, in the sense that *PERSON* behaves similarly to *ANIMAL*, *PHYSICAL OBJECT* to *SOLID TANGIBLE THING*, and *LIVING BEING* to *PERCEPTUAL AGENT*. The problem here, again, comes from incompatible ICs: If a person is a physical object, there is no reason to claim she ceases to exist when she dies, since her body is still there. Indeed, life is considered to be an essential property for a person, while it seems to be an *anti-essential* property for a physical object, in the sense that any physical object (namely, a body) can possibly be a non-living object. If these assumptions are valid, then we must conclude that living beings are not physical objects, but they are rather *constituted* by physical objects.

Example 9 is similar to the previous two, with the difference that here the peculiar IC exhibited by physical devices is a *functional* one rather than a biological one.

Finally, example 10 involves events. The identity conditions for events may be complicated, but it seems plausible to assume as a necessary condition that if two events are the same, then they must have the same participants. Now, the participants involved in physical, mental, and social events are different: We have a physical object, a perceptual agent, and a society, respectively. So the three events are different, although temporally co-located.

6. CONCLUSION

We have presented a compact formalization of some basic issues underlying the notion of identity and unity by assembling, clarifying, and adapting philosophical insights for the purposes of practical knowledge engineering. We believe that the formal meta-properties we have introduced help to make explicit the ontological nature of the concepts used to structure a certain domain, and the constraints they impose on the subsumption relationship force the design of simpler, cleaner, and ultimately more reusable taxonomies.

Unlike previous efforts to clarify the nature of the subsumption relationship, our approach differs in that:

- It focuses on the nature of the properties involved in a single subsumption relationship, not on the semantics of the subsumption relation itself.
- It focuses on the validation of single subsumption relationships based on the *intended meaning* of their arguments in terms of the meta-properties defined here, as opposed to focusing on structural similarities between property descriptions.
- It is founded on formal notions drawn from Ontology, and augmented with practical conceptual design experience, as opposed to being founded solely on the former or latter.

Endnotes

1. Of course, this characteristic relation must be defined for each instance of the class. This means that fingerprints cannot be used as identifying characteristics for persons, since a person may have no fingerprints (while we can assume that each person must have a brain).

2. Note that other trivial identity conditions, such as those discussed in Kaplan (2001) and Carrara & Giaretta (2001), are excluded due to the irreflexivity constraint imposed on χ, which, taking time into account, corresponds to assuming $\forall t \neg \exists x \chi(x,x,t)$. In particular, we believe that this constraint eliminates many trivial instances of $\Sigma \chi(x,y,t,t')$ that imply the identity of x and y.

3. This corresponds to the +O (own identity) mark in previous papers.

4. Consider the following counter example: Suppose you want to say that all the children a, b, c of a certain person form a whole. So all the parts of a+b+c must be linked together by the unifying relation "having the same parent." But two of them, namely a+b and b+c, are not linked by such relation, since they are not persons. Another problem is linked to the fact that the previous definition excludes the possibility of overlapping of entities that are wholes (Kaplan, 2001).

5. See Nirenburg & Raskin (2001) for an objection to this argument, based on a rejection of the role of formal semantics for linguistically-motivated ontologies.

Acknowledgments

We are indebted to Bill Andersen, Stefano Borgo, Massimiliano Carrara, Pierdaniele Giaretta, Dario Maguolo, Claudio Masolo, Chris Partridge, Milena Stefanova, Mike Uschold, and Silvio Valentini for their useful comments on earlier versions of this chapter. This work was supported in part by the Eureka Project (E! 2235) IKF, the Italian National Project TICCA (Tecnologie cognitive per l'interazione e la cooperazione con agenti artificiali), and a research committee grant from Vassar College.

References

Akiba, K. (2000). Identity is simple. *American Philosophical Quarterly*, 37, 389-404.

Brachman, R. (1983). What IS-A is and isn't: An analysis of taxonomic links in semantic networks. *IEEE Computer*, 16(10), 30-36.

Carrara, M., & Giaretta, P. (2001). Identity criteria and sortal concepts. In B. Smith & C. Welty (Eds.), *Proceedings of 2nd International Conference on Formal Ontology in Information Systems (FOIS2001)*, 234-243.

Casati, R., & Varzi, A. C. (1994). *Holes and Other Superficialities*. Cambridge, MA: MIT Press, 1994.

Davidson, D. (1980). The individuation of events. In *Essays on Actions and Events*, 163-180. Oxford: Clarendon Press.

Gangemi, A., Guarino, N., Masolo, C., & Oltramari, A. (2001). Understanding top-level ontological distinctions. In A. Gomez Perez, M. Gruninger, H. Stuckenschmidt, & M. Uschold (Eds.), *Proceedings of Workshop on Ontologies and Information Sharing (IJCAI 2001)*, 26-33.

Goldstein, R. C., & Storey, V. C. (1999). Data abstractions: Why and how? *Data and Knowledge Engineering*, 29, 293-311.

Guarino, N. (1998). Formal ontology and information systems. In N. Guarino (Ed.), *Formal Ontology in Information Systems: Proceedings (FOIS '98)*, 3-15. Amended version available: <http://www.ladseb.pd.cnr.it/infor/Ontology/Papers/FOIS98.pdf> [2001, October 9].

Guarino, N., & Welty, C. (2000a). A formal ontology of properties. *Proceedings of 12th International Conference on Knowledge Engineering and Knowledge Management*, 97-112.

Guarino, N., & Welty, C. (2000b). Identity, unity, and individuality: Towards a formal toolkit for ontological analysis. *Proceedings of the European Conference on Artificial Intelligence (ECAI-2000)*, 211-215.

Guarino, N., & Welty, C. (2000c). *Conceptual Modeling and Ontological Analysis*. AAAI-00 Tutorial Notes. Available: <http://www.cs.vassar.edu/faculty/welty/aaai-2000/> [2001, October 9].

Guarino, N., & Welty, C. (in press). Supporting ontological analysis of taxonomic relationships. *Data and Knowledge Engineering*.

Hirsch, E. (1982). *The Concept of Identity*. New York and Oxford: Oxford University Press.

Hirst, G. (1991). Existence assumptions in knowledge representation. *Artificial Intelligence,* 49, 199-242.

Hughes, G. E., & Cresswell, M. J. (1996). *A New Introduction to Modal Logic.* London: Routledge.

Kaplan, A. (2001). Towards a consistent logical framework for ontological analysis. In B. Smith & C. Welty (Eds.), *Proceedings of 2nd International Conference on Formal Ontology in Information Systems (FOIS2001),* 244-255.

Lewis, D. (1983). New work for a theory of universals. *Australasian Journal of Philosophy,* 61, 343-377.

Lowe, E. J. (1989). *Kinds of Being: A Study of Individuation, Identity and the Logic of Sortal Terms.* Oxford: Basil Blackwell.

Lowe, E. J. (1998). *The Possibility of Metaphysics.* Oxford: Clarendon Press.

Nirenburg, S. & Raskin, V. (2001). Ontological semantics, formal ontology, and ambiguity. In B. Smith & C. Welty (Eds.), *Proceedings of 2nd International Conference on Formal Ontology in Information Systems (FOIS2001),* 151-161.

Noonan, H. (Ed.) (1993). *Identity.* Aldershot, England and Brookfield, VT: Dartmouth.

Simons, P. (1987). *Parts: A Study in Ontology.* Oxford: Clarendon Press.

Storey, V. C. (1993). Understanding semantic relationships. *Very Large Databases Journal,* 2, 455-488.

Strawson, P. F. (1959). *Individuals: An Essay in Descriptive Metaphysics.* London and New York: Routledge.

Chapter 8

Logic of Relationships

Christophe Jouis
Université Paris - Sorbonne Nouvelle, Paris and *CAMS (Centre d'Analyse et de Mathématiques Sociales): CNRS, EHESS, Paris - Sorbonne, Paris, FRANCE*

Abstract:
   A main goal of recent studies in semantics is to integrate into conceptual structures the models of representation used in linguistics, logic, and/or artificial intelligence. A fundamental problem resides in the need to structure knowledge and then to check the validity of constructed representations. We propose associating logical properties with relationships by introducing the relationships into a typed and functional system of specifications. This makes it possible to compare conceptual representations against the relationships established between the concepts. The mandatory condition to validate such a conceptual representation is consistency.
   The semantic system proposed is based on a structured set of semantic primitives—types, relations, and properties—based on a global model of language processing, Applicative and Cognitive Grammar (ACG) (Desclés, 1990), and an extension of this model to terminology (Jouis & Mustafa 1995, 1996, 1997). The ACG postulates three levels of representation of languages, including a *cognitive level*. At this level, the meanings of lexical predicates are represented by semantic cognitive schemes. From this perspective, we propose a set of semantic concepts, which defines an organized system of meanings.
   Relations are part of a specification network based on a general terminological scheme (i.e., a coherent system of meanings of relations). In such a system, a specific relation may be characterized as to its: (1) functional type (the semantic type of arguments of the relation); (2) algebraic properties (reflexivity, symmetry, transitivity, etc.); and (3) combinatorial relations with other entities in the same context (for instance, the part of the text where a concept is defined).

1. WHY IS IT NECESSARY TO INTRODUCE A LOGICAL SYSTEM FOR SEMANTIC RELATIONSHIPS?

   The semantics of the relationships between concepts (i.e., for each relation, the number and types of its arguments, its algebraic properties, etc.) are too often vague (for example, in thesauri, conceptual structures, or semantic networks). The semantics of relationships are vague because the principal users of these relationships are industrial actors (translators of technical handbooks, terminologists, data processing specialists, etc). Nevertheless, the consistency of the models built must always be guaranteed.

   For instance, in terminology, the relationships between concepts are often reduced to the distinction established by standards ISO 704 (1987) and ISO 1087 (1990) between hierarchical relationships (genus-species relationships and part/whole relationships) and non-hierarchical relationships ("time, space, causal relationships, etc.").

One possible approach to this problem consists in organizing the relationships in a typology based on logical properties. For example, Winston, Chaffin, & Herrmann (1987) distinguishes various types of part/whole relationships. This typology inspired the treatment of the part/whole relationship in WordNet (Miller, 1990). Recent work applying terminological relationships to information retrieval, in particular to the construction of thesauri, tries to specify the properties of the link between concepts better and to extend non-hierarchical relationships (Molholt, 1996; Green, 1996, 1998; Bean, 1996). Other recent work aims to integrate into the terminological models theories arising from linguistics (semantics, for example) and artificial intelligence, in particular the modeling of knowledge for the design of knowledge-based systems (KBS) and "ontologies", as defined, for example, by Sowa (1984, 1996, 2000). In all these disciplines, the need to structure knowledge and then to validate the representations obtained is fundamental.

In artificial intelligence, methods for acquiring and modeling knowledge, such as KADS II, as presented in Wielingua, Schreiber, & Breuker (1992), were developed to assist with the design of KBS. These methods propose modeling a field of expertise in the form of concepts connected by semantic relationships in conceptual object-oriented languages (called "domain level" in KADS). From our standpoint, these languages appear very close to terminological database structures.

In terminology, software has been developed to "navigate" networks of concepts structuring micro-domains, for example, the Termisti system (Van Campenhoudt 1994, 1998; Lejeune & Van Campenhoudt, 1998), which considers systems of coherence, the Code system, the Cogniterm project of Meyer and Mchaffie (1994), and the Ikarus system of Meyer and Skuce (1998), which supports computerized management of terminological knowledge bases.

With a view to better designing the knowledge structures underlying the concepts of a field, and more specifically, the indexing of documents and/or information retrieval, we propose a structured set of relationships, based on a linguistic model, the Applicative and Cognitive Grammar (ACG) of Desclés (1990). This model was applied and extended for the acquisition and the modeling of knowledge by Jouis (1993) and then implemented on computer in the SEEK system by Jouis (1998). Mustafa and Jouis (1996, 1997) reconsidered it for the construction of terminologies. The relationships of this model form part of a semantic system for relationships between concepts. We propose a generic schematic of relationships (REL), which is further specified according to algebraic properties in more precise relationships that are axiomatically attributed to them. Our typology is based primarily on the distinction between a field's static situation (its state, situation, "état de choses") and its dynamic situation (modification and change in the field). Note that ours differs from the typology set out by Felbert (1987), who established a distinction between "logical relationships", "ontological relationships", and "relationships of effect". With ours it is possible to check the consistency of the conceptual structure built and to compare it to the relationships established between the concepts.

In the following, we will present the semantic architecture and the four primitive categories of our extension of the ACG linguistic model.

2. PROPOSED SEMANTIC AND LOGICAL SYSTEM

The ACG is an extension of Universal Applicative Grammar (Shaumyan, 1987). It postulates three levels of representation of languages:

- The *phenotype level* (or phenotype) describes such characteristics of natural languages as order of words, morphological cases, etc. *Each language is apprehended in the diversity of linguistic expressions, which are directly observable.* The linguistic expressions of this level are concatenated linguistic units.
- The *genotype level* (or genotype) expresses grammatical invariants and structures that underlie sentences on the phenotype level. The genotype level is structured as a formal language called genotype language. It is described by a grammar called Applicative Grammar (Biskri & Desclés, 1997). Descriptions are represented in the form of applicative expressions formulated with operators and operands of different types.
- In the *cognitive level*, meanings of lexical predicates are represented by semantic cognitive schemes. This level constitutes the knowledge representation associated with one text. Representations of the genotype and cognitive levels are expressions of typed combinatory logic (Curry & Feys, 1958). With the cognitive level, the ACG proposes a set of semantic primitives, which defines an organized system of meanings.

Within the cognitive level, we distinguish four categories of primitive:

- *Elementary semantic types* of entities;
- *Formation operators*, which create more *complex types* from elementary types (lists, arrays, functional types,[1] etc.);
- *Fundamental static relations* between entities, where static relationships enable the description of some states (static situations) related to an area of knowledge and where static situations remain stable during a certain temporal interval where neither the beginning nor the end is taken into account (we have identified over twenty static relationships); and
- *Fundamental dynamic relations* between terminological units, where dynamic relationships enable the description of processes or events related to an area of knowledge: movements, changes of state, conservation of a movement, iterations, intensity, variation, constraints, causes, etc.

Relationships are therefore classified in two main, disjoint categories: static relations and dynamic relations. In this chapter, we will describe more specifically the static relations, because these are the ones we have completely formalized in our system in order to perform coherence checks of the conceptual models. Then, we will give an overview of the dynamic relations.

2.1 Elementary Semantic Types

We distinguish a certain number of elementary types of entities. For instance:

- *Boolean entities* (noted H) are objects, whose value is either true or false.
- *Individualizable entities* are entities that can be designated and shown by pointing. They may be counted individually or regrouped by distributive classes. A quantification operator allows the building of a class more or less determined by individualized

instances. For instance, entities such as *John, table, chair, furniture, man, child* are distinctive. *Individualizable* entities are noted J. For instance: [J: table].

- *Mass entities* such as *water, sea, wine, bread, butter* are not distinctive entities. However, we may notice that a certain number of operators (classificators) can individualize mass notions: *a glass of water, an arm of the sea, a bottle of wine, a slice of bread, a pat of butter.* They are noted M. For instance: [M: sea].

- *Distributive classes* regroup individual entities with one identical property. They are noted D. For instance, [D: to-be-a-square] represents a distributive class of individuals or "concepts".

- *Collective classes*[2] are distinguished from individualizable entities in that they represent objects that form a "whole" from more elementary objects. They are noted C. Thus, [C: geographical entities], [C: army], [C: molecule[3]], [C: human body], etc., represent collective classes.

- *Place*, as a semantic type (noted P), is conceptualized as a set of positions, each position being assimilated to a point. To each entity (individualizable, collective, or mass) we can relate a set of places. For instance, [P: Paris], [P: garden], [P: house], can be on the one hand seen as individualizable entities *(Paris is a city)*, and on the other hand each individualizable entity determines a given *place (I am in Paris).*

2.2 Compound Types

From the set of elementary types S = {H, J, M, D, C, P, . . .}, it is then possible to define a system of more complicated types in a recursive way starting from the following rules:

- The elements of S are elementary types.
- If x and y are types, then Fxy is a (functional) type.

The symbol F is a formation operator of functional types. An entity E of the type Fxy (noted [Fxy: E]) is a unary operator which takes for its argument an object of the type x to provide a result of the type y. If we consider an entity A of type x, the application of E to A will build a certain entity B of the type y:

([Fxy: E] [x: A]) $\rightarrow$ [y: B].

For example, type FJH is that of an operator which, when applied to an individualizable entity (J), returns a value of truth H (unary property of individuals, set of individuals, or "concept", such as [FJH: "to-be-a-square"]).

A relation between an individual entity and a place (localization) will have type FJFPH. Because the localization is a binary operator, the application is done in two steps. For example, the localization of Jean in Paris is formalized in the following way. We have the following types: [J: Jean], [P: Paris] and [FJFPH: localization]. The localization applies initially to Jean to return an operator of the type FPH:

([FJFPH: localization] [J: Jean]) $\rightarrow$ [FPH: localization_Jean]

The result is an operator of the type FPH that applies to the place Paris to return a value of truth v of type H:

([FPH: localization_Jean] [P: Paris]) → [H: True].

The logical connector "AND" is a binary operator which applies to two Boolean entities: [FHFHH: AND], etc.

All representations of the cognitive level are typified in this manner.

2.3 Static Relations Between Entities

The general static relation denoted as REL (or "ε") is a schematic pattern: An entity X is in a relation with an entity Y. This pattern is further specified according to algebraic properties to form more precise relations that are attributed axiomatically to them: identification, differentiation, and disjointedness (see fig. 1).

- The relation of identification (which can be paraphrased as "X is identified with Y", i.e., the entity Y is used as an identifier for entity X) is a binary, symmetric, and reflexive relation. It is expressed in statements such as: *Paris is the capital of France* or *"rendezvous" has the same meaning as "appointment"*.[4]
- The relation of differentiation (which means "X is different from Y") is non-symmetric. We detail this very general relation in Section 2.4.
- The relation of disjointedness (which is read "X is disjoint with Y", i.e., there is no common property between X and Y) is a non-reflexive and symmetric relation. It is expressed in statements like: *Protons and electrons are incompatible.* Very often, this relation operates on disjoint classes derived from the same class by attributions: positive/negative values, quantitative/qualitative properties, etc. Thus the semantic relations form part of a system of meanings of relationships between entities (Culioli & Desclés, 1982; Desclés, 1987; Jouis, 1993).

In this system, a relation may be specified in more precise relations in terms of its properties:[5] (1) its functional type (the semantic type of arguments of the relation), (2) its algebraic properties (reflexivity, symmetry, transitivity, etc.), and (3) its combinatorial relations with other entities in the same context (the part of the text where a concept is defined for instance). For example, inclusion among distributive classes is irreflexive, asymmetric, and transitive. Moreover, in the same context, it is incompatible with some relations, such as the belonging of an individual entity to a distributive class.

Static relations are structured and independent from a particular domain. They are binary relations. We distinguish more than twenty such relations. A number of the properties of the relations of differentiation are explored below.

Among the relations resulting from differentiation, we have first attributions, which are characterized by asymmetry. This asymmetry is specified in:

- The belonging of one individualizable entity (J) to a distributive class (D). Of type FJFDH, this relation is non-reflexive, asymmetric, and non-transitive. It is expressed in statements such as: *π is a real*.[6]
- Inclusion among distributive classes (e.g., *Bacteria are microorganisms*), which is of type FDFDH, is non-reflexive, asymmetric, and transitive.[7]

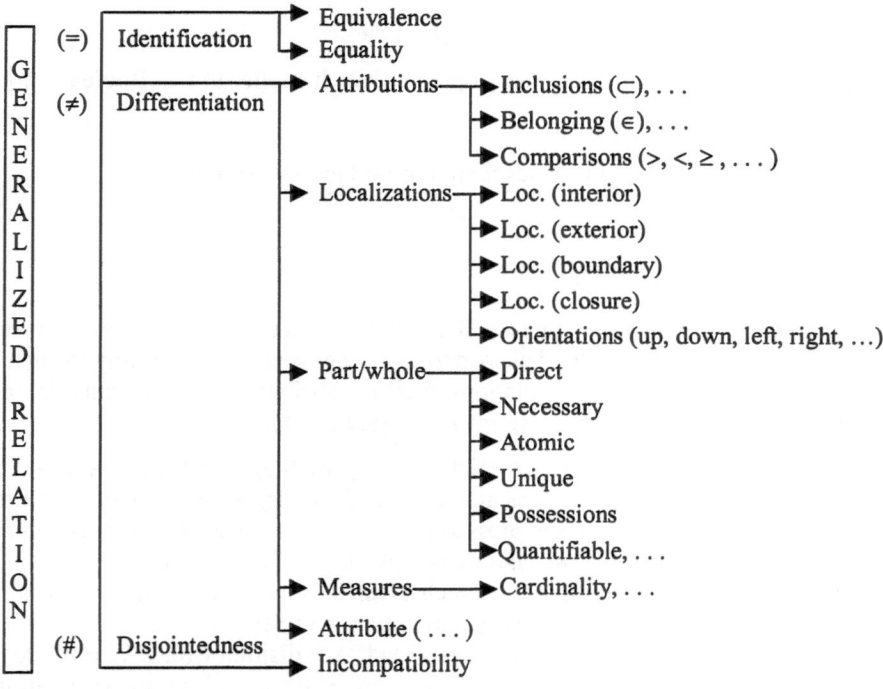

Figure 1. Specification network of static relations

• Comparison, which corresponds to a relation of strict order (i.e., it is neither reflexive nor symmetric, but is transitive), concerning individual entities: Its type is thus FJFJH. It is specified in many relations: superior (>) and inferior (<), etc.

Relations of spatial localization are expressed in the following example statements: *Paris is in France; A garden surrounds the house; The book is on the table*; etc. The location of an entity X (the localized) is related to a place Y (the locator). The relations of localization are of type: FxFPH where x is of type J or type P, according to the context of the entity localized.

Each object occurrence, in a particular pragmatic environment, determines a location (i.e., place or "neighborhood" in the terminology of topology). Primitives of position can be defined by calling upon some basic concepts of general topology. A location is then visualized either in its interior, in its exterior (excluding its interior and its boundary), in its boundary (excluding its interior and its exterior), or as a closure (boundary and interior). We introduce the operators of topological determination of a place x: in (x), ex (x), fr (x) and fe (x), determining the interior, the exterior, the boundary (Fr. *frontière*), and the closure of x (Fr. *fermeture)*, respectively. For any place x, we have, for instance:

$$\text{in } (x) \subset x \subset \text{fe } (x)$$
$$\text{fr } (x) \subset \text{fe } (x) \text{ (because fe } (x) = x \cup \text{fr } (x))$$
$$x \cap \text{ex } (x) = \emptyset$$
$$\text{fr } (x) = \text{co } (\text{in } (x)) \cap \text{co } (\text{ex } (x))$$

The properties of these four operators enable us to establish properties of the localization relations.[8] Specific localization relationships include:

- Loc-in ("to-be-in"): localization in the *interior* of a place (*Oscillating stands in the first zone*). This relation is transitive, asymmetric, and non-reflexive.
- Loc-ex ("to-be-out-of"): localization at the *exterior* of a place (*Limitor is exterior to the third zone*). This relation is not transitive and is never reflexive.[9]
- Loc-ed ("to-be-at-the-boundary"): localization on the *edge-line* (boundary) of a location (*Algiers is at the seaside*). This relation is incompatible with interiority and exteriority; it is more precise than localization at the closure of a locality.
- Loc-cl ("to-be-at-the-closure"): localization at the closure of a locality (*Boulogne is located in the suburb of Paris*). This relation is incompatible with interiority; it is redundant with localization on the *edge-line* and with interiority.

We can distinguish directed localizations in the same way by introducing the primitives left (x), right (x), in front of (x), behind (x), high (x), and low (x). However, these primitives can be defined only if the reference mark object has an intrinsic orientation: the front of the house, the line of a boat (port side), etc.

The part/whole relation is a general relation allowing the decomposition of an object into its components. By using this relation, any individual entity is regarded as an organized complex unit. The part/whole relation admits two arguments, which are, respectively, the broken up object and the component object. Its type is thus FCFxH, where x is of type J or of type C. Part/whole relations include the relations of composition (*Fluorine goes into the composition of bones and teeth*) and possession (*John has got a car*).

The relation of composition is transitive and reflexive, but asymmetric, which differentiates it from inclusion. It is expressed in statements like: *The hand forms part of the arm.* Composition is specified in several relations. Indeed, there are a great number of properties describing the relationship between the composing object and the total object, for example:

- Necessary composition versus non-necessary composition (*The processor is one of the essential components of a computer* versus *A CD-ROM drive is an accessory component of a computer*). The characteristics necessary and non- necessary are transitive within the relation of composition.
- Direct composition versus non-direct composition (*Opium appears among the primary component of Lamaline* versus *A molecule consists of neutrons, protons, and electrons, which are parts of atoms*). An Object-Part OP is a direct component of the Object-Whole OW, if there is no object OP1 (different from OP) such that object OP is a component of object OP1 and object OP1 is a component of object OW. Otherwise, OP is a non-direct component (see fig. 2). Non-direct composition is transitive, while direct composition is not transitive.
- Atomic composition versus non-atomic composition (*The smallest component of a program is the bit* versus *A book breaks up into chapters, which themselves break up into paragraphs*). Atomic composition does not admit transitivity, but non-atomic composition authorizes it.
- Single composition versus non-single composition (*A young star is made up exclusively of atoms of hydrogen* versus *The atmosphere is a mixture of several gases, whose principal ones are oxygen and nitrogen*).

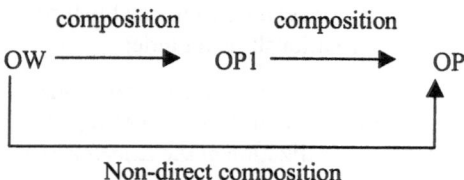

Figure 2. Direct vs. non-direct comparison

- Quantifiable composition versus non-quantifiable composition *(The hand is made up of five fingers; Each human cell contains 46 chromosomes* versus *Water consists of atoms of oxygen and atoms of hydrogen).*

The attribute relation makes it possible to add specific static relations, which are outside the general relation outline, i.e., which are only related to a particular field. For example, the relation "to-be-father-of" that one would use in genealogy cannot be regarded as a general relation. In other words, the structured whole of the relations which we propose is regarded as a whole of semantic invariants, independent of a field considered, but non-exhaustive. Our approach does not exclude the need for adding relations specific to a field.

2.4 Towards Dynamic Relations[10]

In contrast to static relations, dynamic relations build (or describe) non-static situations. They introduce modifications into the field. The modification is a process which passes from a static situation SIT1 to another static situation SIT2. Three temporal zones can be distinguished: (1) before modification (SIT1), (2) during the modification, (3) after modification (SIT2). If we introduce the new elementary types St and Dy for the static situations and dynamic situations, the general outline of a dynamic situation is then:

[St: SIT1] → ([FStFStDy: DYNA]) → [St : SIT2]

Relations noted by the general dynamic relation DYNA describe the transitions.

DYNA is further specified, for example, as relations of space-time movement (MOUVT), relations of change of states (CHANG), relations of conservation of a movement (CONSV), relations of causality (CAUS), etc.

3. TOWARDS A SYSTEM FOR CHECKING THE CONSISTENCY OF CONCEPTUAL STRUCTURES ?

One delicate point is the validation of conceptual structure during its construction. This validation can only be realized with the co-operation of *specialists* (i.e., specialists in the field, terminologists, and/or knowledge engineers).

With relations defined by logico-semantic properties, it is possible to control the consistency of representations by checking that all the properties of the relations are well applied (respected). Consistency is a necessary condition in order to validate conceptual structures. A processing module, testing these properties, is under development in the SEEK system (Jouis, 1993; Jouis & Mustafa, 1995). It integrates specific procedures of control for each property. This module uses GRAPHLET (Himsolt, 1994), a Graph Drawing Processing System. Procedures are released from the moment one introduces a new relation between two concepts. In figure 3, we present an extract of the concepts and relations representation using GRAPHLET.

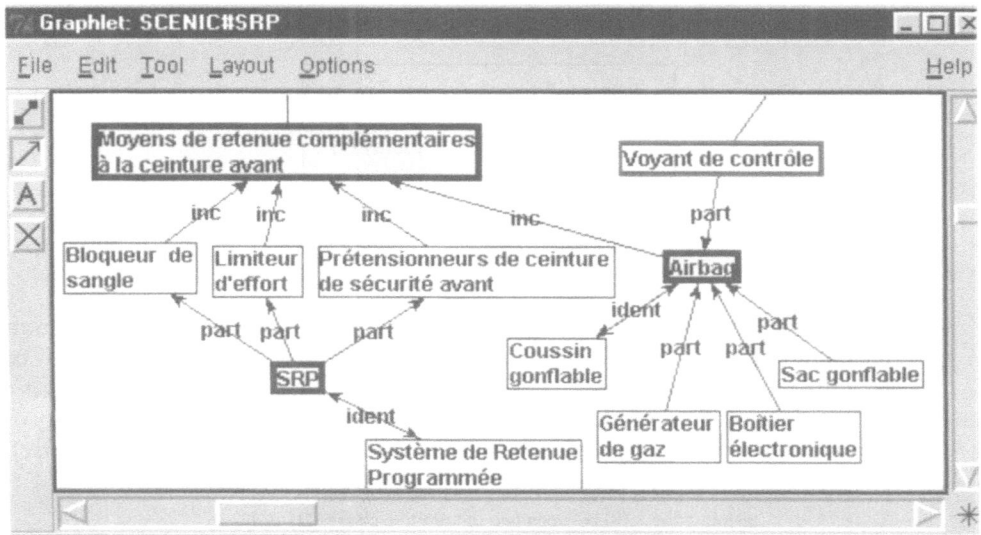

Figure 3. Output of SEEK using GRAPHLET

For example, the relation of inclusion between two distributive classes is not reflexive, is not symmetric, but is transitive. Figure 4 illustrates one of the conditions necessary for the maintenance of this coherency. When one tries to establish a relation of inclusion between a distributive class D1 and another distributive class Dk, it is necessary to verify the non-reflexivity and asymmetry properties, by carrying out transitive closure (route of all the links from Dk, to check that one does not arrive at D1). Let us note that this type of checking can prove to be long and tiresome for the terminologist, who must manage a great number of concepts and relations, especially if it must be done manually.   When a part/whole relation is established between two collective classes D1 and D2, then it becomes incompatible with the relation of inclusion, as represented in figure 5. Let us note that it is necessary to take account of the transitivity of these two relations, which leads to carrying out the transitive closure of these two relations around D1 and D2.

Consistency checks can simultaneously bring into play a great number of properties. Let us consider for example the typical situation given in figure 6. In this example, we have two distributive classes D1 and D2 in the relation of disjointedness (i.e., they are disjoint, disjunction being a specification of the relation of disjointedness). By supposing that the hierarchies of inclusion and membership derived respectively for D1 and D2 are consistent, then, so that the unit remains coherent, it is not possible any more:

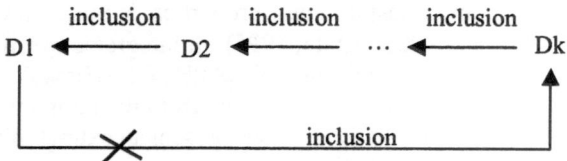

Figure 4. Inclusion is irreflexive, asymmetric, but transitive

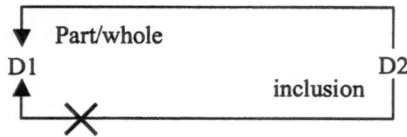

Figure 5. Inclusion and the part/whole relation are incompatible in the same context

- To introduce relations of inclusion between distributive subclasses resulting from the hierarchy of D1 towards distributive subclasses resulting from the hierarchy of D2 (and the contrary); or
- To introduce relations of membership between individual entities of the hierarchy of D1 towards distributive subclasses resulting from the hierarchy of D2 (and the reverse).

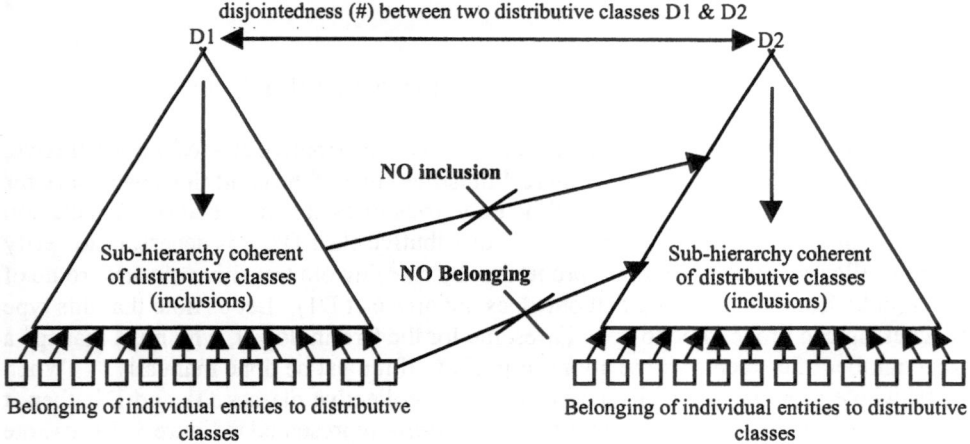

Figure 6. Test of several properties across several relations

These three examples show that the implementation of an information processing system that checks the consistency of a conceptual structure built according to our model consists of simple procedures. These procedures are simple, but must sometimes combine between them. These operations, when they must be carried out manually by the terminologist, require great rigor and long and tiresome work, which our module proposes to realize automatically.

4. CONCLUSIONS

The typology of the relations that we propose is based upon an organized whole of primitives. These primitives are organized starting from semantico-logic types in a logical system of meanings. This system is built gradually, on the basis of a general relational pattern between entities, to get increasingly precise semantic relations by specifying properties progressively.

In this system, the semantics of each relation corresponds to intrinsic properties. By introducing semantic relations into this system, it is then possible to develop an automatic system that checks the internal consistency of conceptual structures (built or added in the course of construction), compared to the properties of the relations. Consistency is a necessary condition (but non-sufficient), with its validation.

The structured set of relations proposed is composed of semantic invariants, independent of any field of knowledge or of any particular language. However, we do not claim that it is complete or exhaustive. The relations should be validated by experimental psycholinguistics.

In addition, our approach does not exclude the need to add relations specific to the field. Moreover, we did not here approach the representation of atypical entities. Lastly, we described only static aspects, to which must be added dynamic relations between terminological units. Such relations make it possible to describe processes or events.

Endnotes

1. In the meaning of the typed lambda-calculus of Church or typed combinatory logic theory of Curry (Curry & Feys, 1958).

2. Lesniewski (1886-1939) proposed a general theory of wholes and parts (mereology), in response to the problems of the paradoxes of set theory (Cantor, 1932/1962; Whitehead & Russell, 1927; Frege, 1964). A detailed analysis of mereology was carried out by Miéville (1984). Regarding a class as an object, a whole is seen as the "accumulation" of elements that constitute it, disjoint or not. Lesniewski (1927/1989) arrives at the conclusion that the notion of class contains two features: the distributive one and the collective one.

The following example, borrowed from Grize (1973, p. 86), gives an idea of the difference: "A distributive class is, to be strictly correct, the extension of a concept. If p is the concept planet, the statement that *Jupiter is a planet* is either to pose pJ or $J \in \{ x \mid px \}$, and the transmitted information is the same one in the two writings. Thus p = { Mercury, Venus, Earth, Mars, Jupiter, Saturn, Uranus, Neptune, Pluto } is a distributive class. It contains nine elements and nothing else. The polar caps of Mars, the red Jupiter spots, the rings of Saturn do not belong to p. Yet all that and a thousand other things deal with the concept planet. The notion of collective class must mitigate this gap."

3. A molecule is formed of different types of atoms, which themselves are formed of . . .

4. The identification is specified in several relations such as extensional equality and intensional equality. On this point, see (Desclés, 1987).

5. For a more complete description of the properties of relations, see Jouis (1993, pp. 146-223). We point out in particular the following properties of reflexivity for binary relations: Given an entity X and a relation R, (completely) reflexive $=_{def}$ $\forall X$ (X R X); nonreflexive $=_{def}$ $\exists X \neg$(X R X); and irreflexive $=_{def}$ $\forall X \neg$(X R X).

6. Consider, however, the attribution of a property to an individual entity. For example, the statement *Socrates is a human* means that the individual entity Socrates belongs to the distributive class of humans or that the concept "to-be-a-man" applies to Socrates.

7. Consider, for example, the statement, *The men are mortals*. It should be noted that, in many thesaurus or semantic network models, we typically use only the general relation "is-a" without distinguishing belonging from inclusion. However, there is a fundamental difference, since the first is not transitive while the second is transitive and allows inheritance of properties.

8. Kuratowski (1958) shows that there are exactly 14 distinct topological operators by combining the four operators identity, in, fe, ex, and complementary co. From this result Barbut (1965) later showed that it is possible to deduce entirely the combinatorial properties of the topological relations of localization. For more details, see Jouis (1993).

9. In other words, $\forall x$, $\neg$(x loc-ex x). This property is even stronger than non-reflexivity, which would give: $\neg(\forall x, (x \text{ loc-ex } x).) \equiv \exists x \neg(x \text{ loc-ex } x)$.

10. For more details, see Abraham (1995).

Acknowledgments

I would like to address my sincere thanks to Catherine Bize (Miami), Olivier Jouis (Nortel Networks Corporation), and Claude Jouis (Ecole Polytechnique / The French Polytechnic School, Palaiseau, France) for their assistance. They enabled me to make this text "consistent" and "coherent" in English.

References

Abraham, M. (1995). *Analyse Sémantico-Cognitive des Verbes de Mouvement et d'Activité: Contributions Méthodologique à la Constitution d'un Dictionnaire Informatique des Verbes*. Doctoral dissertation, Ecole des Hautes Etudes en Sciences Sociales.

Barbut, M. (1965). Topologie générale et algèbre de Kurarowski. *Mathématiques et Sciences Humaines,* 12, 11-27.

Bean, C. (1996). Analysis of non-hierarchical associative relationships among Medical Subject Headings (MeSH): Anatomical and related terminology. *Knowledge Organization and Change: Proceedings of the Fourth International ISKO Conference,* 80-86.

Biskri, I., & Desclés, J.-P. (1997). *Applicative and Combinatorial Grammar: From Syntax to Functional Semantics.* Amsterdam and Philadelphia: John Benjamins.

Cantor, G. (1962). *Gesammelte Abhandlungen* (E. Zermelo, Ed.). Hildesheim: G. Olms. (Original work published 1932.)

Culioli, A., & Desclés, J.-P. (1982). Traitement formel des langues naturelles. *Mathématiques et Sciences Humaines,* 77, 3-125; 78, 5-31.

Curry, H., & Feys, R. (1958). *Combinatory Logic* (Volume I). Amsterdam: North-Holland.

Desclés, J.-P. (1990). *Langages Applicatifs, Langues Naturelles et Cognition.* Paris: Hermes.

Desclés J.-P. (1987). Réseaux sémantiques: La nature logique et linguistique des relateurs. *Langages,* 87, 55-78.

Felber, H. (1987). *Manuel de Terminologie.* Paris: UNESCO.

Frege, G. (1964). *Begriffsschrift und andere Aufsätze* (I. Angelelli, Ed.; 2nd ed.). Hildesheim: G. Olms.

Green, R. (1998). Attribution and relationality. *Structures and Relations in Knowledge Organization: Proceedings of the Fifth International ISKO Conference,* 328-335.

Green, R. (1996). Development of a relational thesaurus. *Knowledge Organization and Change: Proceedings of the Fourth International ISKO Conference,* 72-79.

Grize, J.-B. (1973). *Logique Moderne* (Fascicule II). Paris: Mouton/Gauthier-Villars.

Himsolt, M. (1994). GraphEd: A graphical platform for the implementation of graph algorithms. In R. Tamassia & I. G. Tollis (Eds.), *Graph Drawing, Lecture Notes in Computer Science 894,* 182-193.

International Organization for Standardization (ISO). (1990). *Terminology - Vocabulary = Terminologie - Vocabulaire.* Genève: Organisation internationale de normalisation. (ISO 1087-1990.)

International Organization for Standardization (ISO). (1987). *Principes et Méthodes de la Terminologie.* Genève: Organisation internationale de normalisation. (ISO 704-1987).

Jouis, C. (1998). System of types + inter-concept relations properties: Towards validation of constructed terminologies? *Structures and Relations in Knowledge Organization: Proceedings of the Fifth International ISKO Conference,* 39-47.

Jouis, C., & Mustafa-Elhadi, W. (1996). Vers un nouvel outil interactif d'aide à la conception de dictionnaires électroniques spécialisés. *Lexicomatique et Dictionnairiques: IVes Journées Scientifiques du Réseau Thématique "Lexicologie, Terminologie, Traduction",* 255-266. Beyrouth: AUPELF-UREF & F.M.A.

Jouis, C., & Mustafa, W. (1995). Conceptual modeling of database sketch using linguistic knowledge: Application to terminological databases. *Proceedings of the First Workshop on Applications of Natural Language to Data Bases,* 103-118.

Jouis, C. (1993). *Contributions à la Conceptualisation et à la Modélisation des Connaissances à partir d'une Analyse Linguistique de Textes: Réalisation d'un Prototype: Le Système SEEK.* Doctoral dissertation, Ecole des Hautes Etudes en Sciences Sociales.

Kuratowski, K. (1958). *Topologie* (4th ed.). Warszawa: Panstwowe Wydawn. Naukowe.

Lejeune, N., & Van Campenhoudt, M. (1998). Modèle de données et validité structurelle des fiches terminologiques: L'expérience des microglossaires de TERMISTI. In Blanchon, (Ed.), *La Banque des Mots: Terminologie et Qualité,* numéro spécial 8, 97-111.

Lesniewski, S. (1989). Sur les fondements de la mathématique: Fragments (discussions préalables, méréologie, ontologie) (G. Kalinowski, Trans.). Paris: Hermés. (Original work published 1927.)

Meyer, I., & Skuce, (1998). Bases de connaissances et bases textuelles sur le web: Le système Ikarus. In A. Clas, S. Mejri, & T. Baccouche (Eds.), *Vèmes Journées Scientifiques: La Mémoire des Mots,* 637-646. Tunis and Montréal: AUPELF & F.M.A.

Meyer, I., & Mchaffie, C. B. (1994). De la focalisation à l'amplication: Nouvelles perspectives de représentation des données terminologiques. In A. Clas & P. Bouillon, (Eds.), *T.A.-T.A.O.: Recherches de Pointe et Applications Immédiates: Troisièmes Journées Scientifiques du Réseau Thématique de Recherche "Lexicologie, Terminologie et Traduction",* 425-440. Montréal:AUPELF-UREF & F.M.A.

Miéville, D. (1984). *Un Développement des Systèmes Logiques de Stanislaw Lesniewski: Protothétique, Ontologie, Méréologie.* Bern, Frankfurt am Main, and New York: P. Lang.

Miller, G. A. (1990). Nouns in WordNet: A lexical inheritance system. *International Journal of Lexicography.* 3, 245-264.

Molholt, P. (1996). Standardization of interconcept links and their usage. *Knowledge Organization and Change: Proceedings of the Fourth International ISKO Conference,* 65-71.

Mustafa-Elhadi, W., & Jouis, C. (1996). Evaluating natural language processing systems as a tool for building terminological databases. *Knowledge Organization and Change: Proceedings of the Fourth International ISKO Conference,* 346-355.

Mustafa, W., & Jouis, C. (1997). Natural language processing-based techniques and their use in data modelling and information retrieval. *Knowledge Organization for Information Retrieval: Proceedings of the Sixth International Study Conference on Classification Research,* 157-161. The Hague: FID.

Shaumyan, S. (1987). *A Semiotic Theory of Language.* Bloomington: Indiana Univ. Press.

Sowa, J. F. (1984). *Conceptual Structures: Information Processing In Mind And Machine.* Reading, Mass.: Addison-Wesley.

Sowa, J. F. (1996, August). Ontologies for knowledge sharing. Paper presented at Terminology and Knowledge Engineering '96) Vienna, Austria.

Sowa, J. F. (2000). *Knowledge Representation: Logical, Philosophical, and Computational Foundations.* Pacific Grove, CA: Brooks/Cole.

Van Campenhoudt, M. (1998). Abrégé de Terminologie Multilingue [On-line]. Available: <http://www.termisti.refer.org/marcweb.htm> [2001, October 9].

Van Campenhoudt, M. (1994). Les relations notionnelles expérimentées dans les microglossaires de TERMISTI: Du foisonnement à la régularité. In A. Clas & P. Bouillon, (Eds.), *T.A.-T.A.O.: Recherches de Pointe et Applications Immédiates: Troisièmes Journées Scientifiques du Réseau Thématique de Recherche "Lexicologie, Terminologie et Traduction",* 409-423. Montréal and Beyrouth : AUPELF-UREF & F.M.A.

Whitehead, A. N., & Russell, B. (1927). *Principia Mathematica* (2nd ed.). Cambridge and New York: Cambridge University Press.

Wielinga, B., Schreiber, A., & Breuker, J. (1992). KADS: A modeling approach to knowledge engineering. *Knowledge Acquisition,* 4 (1), 5-53.

Winston, M. E., Chaffin, R., & Herrmann, D. (1987). A taxonomy of part-whole relations. *Cognitive Science,* 11, 417-444.

PART III

Applications of Relationships

Chapter 9

Thesaural Relations in Information Retrieval

Martha Evens
Department of Computer Science, Illinois Institute of Technology, Chicago, IL, USA

Abstract:
   Thesaural relations have long been used in information retrieval to enrich queries; they have sometimes been used to cluster documents as well. Sometimes the first query to an information retrieval system yields no results at all, or, what can be even more disconcerting, many thousands of hits. One solution is to rephrase the query, improving the choice of query terms by using related terms of different types. A collection of related terms is often called a thesaurus. This chapter describes the lexical-semantic relations that have been used in building thesauri and summarizes some of the effects of using these relational thesauri in information retrieval experiments.

1. INTRODUCTION

   This chapter will explore some of the ways that thesauri have been used in information retrieval, the different types of thesauri that have been built, and the use of particular kinds of thesaural relations in these experiments. Then we will look at the history of the use of relational models in information retrieval, beginning with early experiments by Sparck Jones and others in Great Britain and by Salton and his students using his SMART system in the United States. About twenty years ago independent work by Edward Fox and Michael Lesk led to an important series of experiments with machine readable dictionaries. While the early experiments performed in batch mode on main frames tended to clump all relations together, the advent of personal computers and interactive retrieval systems made it feasible to differentiate relations. This led to increasing emphasis on new inventories of relations developed by computational lexicographers like Roy Byrd at IBM, Nicoletta Calzolari at Pisa, and others. New sets of relations also grew out of the WordNet project, and the development of the Unified Medical Language System (UMLS) by the National Library of Medicine.
   In information retrieval, thesauri have been used most often to enhance queries by adding words and phrases, but they can also be used to augment the text of the documents. This augmented text can then be used to index the documents, to cluster them, and to build networks of related documents. In the most extreme case, the actual document text has been replaced by the thesaurus terms during the retrieval process (Voorhees, 1998).
   Since thesauri are typically large and complex constructs themselves, a major factor in the use of thesauri has always been their availability, or the availability of knowledge bases and tools for thesaurus construction. We will look briefly at available thesauri and at thesaurus construction techniques as we go.

Information retrieval has changed markedly from its beginnings in the days of batch processing on machines with storage for citations for a few hundred documents, through the arrival of personal computers and interactive feedback, to today's systems for searching full text among the gigabytes available on the Internet. The Tipster project of the Defense Advanced Research Projects Agency (DARPA) has played a major role in these changes, directing the attention of the research community from academic documents with bibliographies to newspaper articles of different lengths and mandating software trials involving hundreds of thousands of documents instead of a few hundred or a few thousand (see the Tipster Project reports published in the Text Retrieval Conference (TREC) Proceedings by Harman, 1993-). Due partly to the emphasis placed by the Tipster project on multilingual retrieval approaches, similar experiments have been carried out on a smaller scale in languages as widely divergent as Japanese, Chinese, Finnish, Arabic, and Italian. These changes have had a strong impact on both the ways that thesauri are used and on their construction.

We wind up with a brief discussion of the work with thesauri reported at TREC and with some speculations on the future of lexical and semantic relations in information retrieval as "intelligent" approaches make more use of natural language processing techniques.

2. ROGET'S THESAURUS

The best known thesaurus in the English-speaking world is *Roget's International Thesaurus*. This thesaurus was compiled as a reference for writers. It provides lists of associated words, called synonyms, along with a structure to help the user to discriminate among word senses. *Roget's Thesaurus* served as the basis for a series of early information retrieval experiments carried out by Sally and Walter Sedelow and their students (Sedelow, 1973, 1991).

Roy Byrd and his colleagues at IBM (Byrd et al., 1987) also used *Roget's Thesaurus* in the development of a large thesaurus for use in disambiguating word senses for information retrieval and other natural language processing tasks. In addition to Roget they used the machine-readable versions of the *Collins Thesaurus* and of *Webster's Seventh Collegiate Dictionary* and *Longman's Dictionary of Contemporary English*. From Byrd's (1989) description of this project, the relationships are derived mainly from the defining formulae in the dictionary definitions, while the thesauri were used to make sure that all the related words were included.

3. RELATIONAL MODELS IN THE SOCIAL SCIENCES

Converging research on relational models in anthropology, psychology, and linguistics has also been a fruitful source of thesaurus relations. The relational models of Casagrande and Hale (1967), Russell and Jenkins (1954), and Apresyan, Mel'cuk, and Zholkovsky (1970) have had the most direct influence on research in information retrieval.

Casagrande and Hale (1967) began by collecting 800 folk definitions in Pima and Papago and extracting relationships between words and phrases. They identified thirteen different defining relations from their data (the first thirteen relations in table 1). They then turned to the Minnesota Norms of Russell and Jenkins (Russell & Jenkins, 1954; Jenkins, 1970) and found all thirteen of their original relations plus four more (also shown in table 1). We have replaced their original examples by computer examples that we thought might be more understandable to our readers. The relations are listed in order of frequency in the corpus of 800 definitions. Note that attributive and functional relationships were used much more frequently than class inclusion (taxonomy or ISA) or synonymy, the main defining strategies in commercial dictionaries today. These dictionaries reflect the Aristotelian teaching that a definition (especially a noun definition) should contain a genus term and differentiae.

Relation	Definition	Example
Attributive	defined in terms of an important attribute	a monitor has a screen
Contingency	defined in terms of a concomitant action	when the electricity goes off the computer crashes
Function	defined in terms of an action of which it is a characteristic instrument	a search engine is how we find documents
Spatial	defined in terms of spatial relationships	a button is an area of a screen
Operational	defined in terms of an action of which it is a characteristic goal	a button is what we click on
Comparison	defined by comparison with a similar object	a palmtop is like a laptop, but smaller
Exemplification	defined in terms of a typical exemplar (e.g., green like grass)	small, the size of a palmtop
Class inclusion	defined by hyponymy; the ISA relation	a main-frame is a computer
Synonymy	defined by giving an equivalent term	Web-based is the same as Internet-based
Antonymy	opposite (used primarily with adjectives)	user-friendly means not annoying to use
Provenience	source or composition	a chip is made of silicon
Grading	adjacent in a generic list	kilobytes, megabytes, gigabytes; private, corporal, sergeant
Circularity	a word is defined in terms of morphologically related words	a computable function is one that can be computed
Constituent	defined in terms of parts or wholes	a CPU is part of a computer
Coordinate	words often coordinated	client-server
Clang	rhyming words	table-stable
Sequential	words that commonly cooccur; another syntagmatic relation, like the coordinate relation	operating system neural network

Table 1. Relations of Casagrande and Hale (1967)
(The examples in this table were created by the author.)

Experiments in collecting folk definitions in other languages, particularly English, have revealed very similar sets of relations. Like the work of Casagrande and Hale, they also show that taxonomy and synonymy are used much less often than attributive, functional, and operational strategies in the construction of folk definitions.

At about the same time, Soviet linguists began to develop relational models of the lexicon. Mel'cuk's *Explanatory-Combinatorial Dictionary* (ECD) (Mel'cuk & Zholkovsky, 1988) was central to this effort. The ECD relations appear, among others, in table 2.

GROUP Lexical relation	Argument	Related word	Explanation
A. CLASSICAL			
Synonymy	amusing	funny	substitutable word (M Syn)
Taxonomy	lion	animal	ISA (M Gener)
B. ANTONYMY			
Comp	married	single	binary opposites
Anti	hot	cold	one denies the other (M)
Conv	to buy	to sell	conversiveness (M)
Reck	husband	wife	reciprocal kinship
C. GRADING			
Queue	Monday	Tuesday	adjacent in a list
Set	sheep	flock	aggregate (M Mult)
Stage	ice	water	manifestation of (M Manif)
Compare	wolf	coyote	typically compared with
D. ATTRIBUTE			
Male	duck	drake	unmarked to male
Female	lion	lioness	unmarked to female
Child	cow	calf	parent to juvenile
Home	lion	Africa	origin, habitat
Son	dog	bark	characteristic sound (M)
Madeof	tire	rubber	substance, made of
Color	tomato	red	usual color
Time	breakfast	morning	usual time
Location	toilet	bathroom	usual place
Size	giraffe	tall	usual height or volume
Quality	saint	holy	characteristic attribute (M Quali)
E. PARTS-WHOLES			
Part	tusk	elephant	has part
Cap	tribe	chief	head of organization (M)
Equip	gun	crew	staff or personnel (M)
Piece	sugar	lump	item of (M Sing)
Comes-from	milk	cow	provenience
Poss	rich man	money	(QUAL1) possesses

Table 2. Groups of lexical relations used by Fox (1981)
(If the relation comes from the ECD, the explanation column contains an M for Mel'cuk.)

GROUP Lexical relation	Argument	Related word	Explanation
F. CASE			
Tagent	conquer	conqueror	typical agent
Tobject	dine	dinner	typical direct object (M S2)
Tresult	dig	hole	result (M Sres)
Tcagent	beat	loser	counter-agent
Tinst	sew	needle	instrument (M Sinstr)
Tsource	sprout	earth	typical source
Texper	love	lover	experiencer
Tloc	bake	kitchen	location (M Sloc)
Tsubject	sell	seller	subject (M S1)
Tinobj	sell	price	indirect object (M S4)
G. PREDICATES			
Perm	to drop	to fall	permit, make possible (M)
Incep	difficulty	to run into	to begin (M)
Cont	peace	to maintain	to continue (M)
Fin	patience	to lose	to cease or stop (M)
Perf	study	to master	to accomplish (M)
Result	dead	to die	to become (M S0)
Fact	dream	to come true	to become a fact (M)
Real	to attempt	to succeed	to make a fact (M)
H. COLLOCATION			
Copul	victim	to fall	the copula verb (M)
Liqu	mistake	to correct	the destroy verb (M)
Prepar	table	to set	the prepare verb (M)
Degrad	teeth	to decay	to deteriorate (M)
Inc	tension	to mount	the increase verb (M)
Dec	cloth	to shrink	the decrease verb (M)
Bon	conditions	favorable	positive adjective (M)
Ver	reason	valid	proper adjective (M)
Centr	life	prime	culmination (M)
I. MORPHOLOGY			
Past	go	went	past tense form
PP	go	gone	past participle form
Plural	man	men	irregularplural form
Others	fun	funny	derivational morphology (M A1)

Table 2. Groups of lexical relations used by Fox (1981)—Cont.
(If the relation comes from the ECD, the explanation column contains an M for Mel'cuk.)

GROUP Lexical relation	Argument	Related word	Explanation
J. PARADIGMATIC			
Cause	to send	to go	cause and effect (M Caus)
Become	red	to redden	verb to get that result (M)
Be	near	to neighbor	that which is (M Pred)
Nomv	death	to die	process noun-verb (M inv S0)
Adj	sun	solar	adjective form (M inv A0)
Able	burn	combustible	able to (be) (M Ablei)
Imper	talk	go ahead	irregular imperative (M)
Per	cold	freezing	intensified form
Mode	style	to write	mode of action (M Smod)
Figur	passion	flame	figurative designation
K. SITUATIONAL			
Ai	fire	to burn	generic attribute (M Vi)
Operi	sacrifice	to make	situational verb (M Operi)
Funci	silence	to reign	functional verb (M Funci)
Labori	torture	to put to	verb for action (M Labori)
Si	sell	goods	name of participant (Si)
Magn	rain	heavy	name of intensifier (M)

Table 2. Groups of lexical relations used by Fox (1981)—Cont.
(If the relation comes from the ECD, the explanation column contains an M for Mel'cuk.)

4. SALTON AND HIS STUDENTS

Gerard Salton emerged as a major figure in the new science of information retrieval in the 1950's and remained its leader until his death forty years later. He built a test-bed called the Smart system and used it to carry out experiments on every new retrieval technique that came along including the construction and use of thesauri. In order to understand his experiments, we need to look at his definition of "thesaurus". Salton describes a thesaurus this way: "A thesaurus, or synonym dictionary, which specifies for each dictionary entry one or more synonym categories or concept classes . . . a thesaurus is then used to perform a many-to-many mapping from word entries to concept classes" (1968, p. 25). In other words, "a thesaurus is a grouping of words, or word stems, into certain subject categories, hereafter called concept classes" (p. 25). We can see that Salton was interested in only one thesaurus relation, synonymy, and his main goal for the thesaurus was "consistent language normalization" (p. 23). His approach becomes even clearer, if we follow his construction of an example thesaurus class (pp. 26-29).

Composing classes out of words that frequently co-occur gives for Concept Class #101 (CALL) the list: DESIGNATE, IDENTIFY, IDENTIFIER, IDENTIFICATION, INDEX, INDICATE, LABEL, MARK, NAME, POINT, SIGNAL, SIGN, SUBSCRIPT, and TAG. Next these words are stemmed and *identify, identifier,* and *identification* are combined into one entry. Now it is time to apply Salton's five basic principles of thesaurus construction. First and foremost, very rare concepts should not be included because they do not have enough matches in the document collection to give useful results. Second, very common high frequency terms should be eliminated because they produce too many matches. Studying a list of word frequencies in the particular document collection, he decided to remove INDICATE, CALL, NAME, and DESIGNATE as too common to be useful index terms. Third, Salton advises, word senses need to be considered carefully. The words in the IDENTI family were moved into a separate concept class representing the concept of recognition. Fourth, it is important to include only the relevant senses of ambiguous words. Thus, SIGNAL and SIGN were removed because, in this particular document collection, they appeared only in the context of pulse signals, not in the sense of using names to represent concepts. Fifth, it is important to include concepts of roughly equal frequency. The final list of terms in concept class #101 is: INDEX, LABEL, SUBSCRIPT, and TAG. A carefully controlled document retrieval study looking at both recall and precision showed that a statistical thesaurus constructed in this manner gave better results than a hand-built thesaurus (1968, p. 44).

Salton made it clear that the goals of the application can and should affect the choices made in thesaurus construction. He pointed out the tradeoffs involved in choosing "broad, inclusive concept classes" (which improve recall at the cost of precision) vs. "narrow and specific classes" (which improve precision but depress recall) (1968, p. 27).

In addition to the "regular thesaurus", which "provides synonym recognition", Salton also discussed the use of a "concept hierarchy" to enrich queries. This concept hierarchy is essentially an ontology structured by ISA relations (1968, p. 44). Salton experimented with more automatic methods of thesaurus construction, based on a more detailed statistical analysis, and also on automatic methods of hierarchy construction, but he continued to base his work on only two relations, synonymy and taxonomy.

5. THESAURUS CONSTRUCTION BASED ON SEMANTICS: KAREN SPARCK JONES

One of the first to carry out systematic experiments using a thesaurus was Karen Sparck Jones (1971). She was convinced that a useful thesaurus had to be based on an automatic classification system, but she wanted to find an alternative to pure statistical cooccurrence, so she developed her own theory of semantic classification. The basic element in Sparck Jones's theory of semantic classification is a *run,* a set of word forms that are interchangeable in some context. Since her goal was to solve information retrieval problems, this definition is based on strings instead of senses. So *knot* and *greeting* are in the same run, since both can substitute for *bow* in some (different) contexts. Since, fortunately, this degree of polysemy is rather rare, runs tended to include synonyms and hypernyms and hyponyms.

Sparck Jones (1971) found that using this kind of automatic keyword classification in information retrieval experiments was generally effective. Minker, Wilson, and Zimmerman (1972) failed to find the same effect, but Salton (1973) questioned Minker's methodology and carried out confirming experiments of his own.

6. EDWARD FOX AND NEW INVENTORIES OF RELATIONS

Ten years later, Edward Fox (1981), while still a graduate student working with Salton at Cornell, decided that it was important to consider using a wider range of relations in retrieval. He learned of Mel'cuk's Meaning-Text Model, decided it was perfectly suited to work in information retrieval, and designed a thesaurus using Mel'cuk's relations (Apresyan, Mel'cuk, & Zholkovsky, 1970). Given the large number of relations in Mel'cuk's model and the complexity of carrying out experiments using different thesauri in the Smart system, Fox decided to use the groups of relations developed by Evens and Smith (1978). These groups, shown in table 2, include Mel'cuk's relations and some others used in anthropology (Berlin, Breedlove, & Raven, 1966; Casagrande & Hale, 1967) and in psychology (Riegel, 1970; Norman, 1976).

The morphology relations are certainly paradigmatic relations (relations between words that express the same semantic core) and these two categories should probably have been merged. Many of the relations are, in fact, syntagmatic or collocational (they link words that appear together in a sentence), especially the predicates and the situational relations, so it might make sense to combine these categories as well.

Fox found that enriching queries using a thesaurus based on each one of these groups separately produced significant improvement, except for the group of relations labeled antonymy. He obtained even better results combining all the groups together, but leaving out the antonymy relations. At the close of his paper he suggests that some of the advantage from pooling all the relations may be an artifact of the batch-processing environment and speculates about the possibility of using these relations in an interactive environment.

Wang, Vandendorpe, and Evens (1985) repeated these experiments using the same groups of relations but a different information retrieval system and a different document collection. They obtained very similar results showing the advantages of using a relational thesaurus to enrich queries in an established collection of documents. Again, the use of the antonymy relation group made the results worse, and the combination of all the relations except antonymy gave the best results.

Salton criticized these efforts rather severely, arguing that the construction of semantic thesauri was too labor intensive and that the best approach was the use of statistical thesauri, based principally on word co-occurrence, since statistical thesauri can be built using fully automatic methods of thesaurus construction. He was willing to grant that thesauri based on semantics were potentially just as valuable, perhaps even more so, but since they could not be constructed automatically, such thesauri were not feasible as a basis for large-scale information retrieval experiments.

7. COMPUTER-AIDED THESAURUS CONSTRUCTION

During the mid 1970's several British dictionary publishers made machine-readable dictionaries available for research for the first time. Since such dictionaries have a large amount of specific semantic information, it seemed feasible to try to find automatic ways of building semantic thesauri starting from a machine-readable dictionary.

Ever since he published his analysis of the structure of the *Merriam-Webster Pocket Dictionary* (Amsler, 1981), Robert Amsler has been the leading figure in research on machine-readable dictionaries. Walker and Amsler (1986) used the resources Amsler developed in a series of information retrieval experiments.

In the mid 1980's Edward Fox joined forces with Evens in an experiment aimed at building a large thesaurus for information retrieval automatically from the machine-readable version of *Webster's Seventh Collegiate Dictionary*. The first emphasis here was on developing an inventory of defining formulae like "to make or become" and "of or relating to" that could be mapped into specific lexical-semantic relations automatically (Ahlswede, 1985; Ahlswede & Evens, 1988a, 1988b; Nutter, Fox, & Evens, 1990).

When the thesauri constructed in this way were used in information retrieval experiments, the results were equivocal. When the vocabulary involved was restricted to a particular sublanguage and ambiguities were weeded out, large improvements appeared in retrieval experiments (Ahlswede, Evens, Markowitz, & Rossi, 1986; Ahlswede et al., 1988; Fox, Nutter, Ahlswede, Evens, & Markowitz, 1988). When a large vocabulary was used without reference to the underlying sublanguage and lists of related words appearing in dictionary definitions were added to the query without any attempts at disambiguation (Wang, Evens, & Hier, 1993), the improvement in recall was swamped by the degradation in precision. It is clear that truly automatic thesaurus construction must include sense disambiguation. Ahlswede has done further work on the disambiguation of dictionary references. In particular, he has shown what a difficult task this can be for human beings (Ahlswede, 1992; Ahlswede & Lorand, 1993; Ahlswede, 1995). Future SENSEVAL Conferences may provide significant help in this area.

Michael Lesk had some remarkable success in this area using the *Oxford English Dictionary*, which has longer and more detailed definition text than any other dictionary of the English Language. By using a complex algorithm for counting relationships between words in the definition text and words in the semantic field of a particular dictionary sense, he managed to discriminate between *pine cones* and *ice-cream cones,* for instance (Lesk, 1987). He later carried out a number of other experiments with the OED, but none were sufficiently large-scale to be overwhelmingly convincing.

Calzolari (1988) argued that the "dictionary and the thesaurus can be combined" and should be. For a number of years Calzolari has been the director of the DMI or Italian Machine Dictionary project at the University of Pisa. She has built a magnificent lexical database of over one million word forms from 120,000 roots. This database has been used for parsing, generation, and machine translation projects as well as information retrieval. Noun entries originally contained pointers to synonyms and antonyms. Extensive work on building ontologies made it possible to add hypernym and hyponym pointers as well. Calzolari's next step was an analysis of typical case relationships in Italian and the addition of this information to verb entries. Techniques developed in this work were then used to develop explicit information about a wide range of lexical-semantic relations (Calzolari & Picchi, 1988) including ACT, BECOME, INHABITANT, INSTRUMENT, LIQUID, LOSE, PART, PLACE, SCIENCE, SET, SOUND, and STUDY, among others. More recently, she has analyzed the semantics of various kinds of derivational morphology and recorded these morphological relationships explicitly in the dictionary. Pentheroudakis and Vanderwende (1993) developed algorithms for this kind of analysis in English using

LDOCE. Calzolari is currently playing a leading role in the EEC project to develop parallel multilingual resources for use in machine translation and in cross-language information retrieval (Calzolari, Baker, & Kruyt, 1996; Ruimy et al., 1998).

8. WORDNET RELATIONS AND WORDNET-BASED IR EXPERIMENTS

WordNet is the creation of the psycholinguist, George Miller, who set out to build a model of the lexical information stored in human memory (Miller, Beckwith, Fellbaum, Gross, & Miller, 1991). Miller starts off his discussion of relations in WordNet by telling us that the basic semantic relation is synonymy. The nodes in WordNet are *synsets* or sets of synonymous word senses. He tells us to "think of a synset as representing a lexicalized concept of English" (1998, p. 24). But, he goes on, "the semantic relation that is most important in organizing nouns . . . is the relation subordination (or class inclusion or subsumption)," which he calls *hyponymy*. When work on WordNet began, the plan was to use only three relations, *hyponymy, meronymy,* and *antonymy*. As WordNet grew, this inventory of relations was not enough to characterize all the differentiae. At this point Miller and his colleagues decided not to add more relations but instead to add parenthetical glosses to differentiate between senses of polysemous words (Miller, 1998, pp. 36-37). Thus one sense of the verb "eat" is glossed "take in solid food". These glosses can also be used in information retrieval.

Note that synsets are very different from the runs developed and used by Sparck Jones. Runs are defined in terms of character strings. Thus the run for "bow" contains longbows and front ends of ships and gestures of respect and greeting, but these concepts are associated with different word senses, so they all belong to separate synsets (Fellbaum, 1998).

One of the first to use WordNet in information retrieval was Ellen Voorhees (1998). She carried out two series of experiments. In the first study WordNet synsets were used to represent the contents of the documents. In the second WordNet was used to augment user queries. Voorhees cited the query in table 3 (TREC query #122) to explain her results.

Voorhees's experiments involved five standard test collections: CACM, CISO, CRAN, MED, and TIME. In the first study Voorhees compared the retrieval results using stemmed nouns with retrieval results using synsets. Using synsets made the results worse. She explains: "Examination of the individual query results shows that most of this degradation is caused by term matches between documents and queries that are made in the standard run but missed in the sense-based runs" (1998, p. 289). She concludes, "This analysis indicates that specialization/generalization relations are unlikely to contain sufficient information to choose among fine sense distinctions" (p. 294).

Voorhees (1994) reports on two sets of experiments to explore the possibility of using WordNet for query expansion. The first set used hand-picked synsets. The second used automatic methods of choosing the synsets.

Voorhees picked the synsets {*cancer*}, {*skin-cancer*} and {*pharmaceutical*} to enhance TREC query #122, shown in table 3. Once the synsets are chosen, there are several ways to use them. Voorhees experimented with four different strategies: expanding the query with the words in the synset, expanding the query with the words in the synset and all

<dom> Domain	Medical & Biological
<title> Topic	RDT&E of New Cancer Fighting Drugs
<desc> Description	Document will report on the research, development, testing and evaluation (RDT&E) of a new anti-cancer drug developed anywhere in the world.
<narr> Narrative	A relevant document will report on any phase in the worldwide process of bringing new cancer fighting drugs to market, from conceptualization of government marketing approval. The laboratory or company responsible for the drug project, the specific type of cancer(s) which the drug is designed to counter, and the chemical/medical properties of the drug must be identified.
<con> Concept(s)	1. cancer, leukemia 2. drug, chemotherapy

Table 3. TREC query 122 from Voorhees (1998, p. 296)
(The summary statement is in the description field)

descendants in the ISA hierarchy, expanding the query with the words in the synset and all their descendants and also their parents, and expanding the query with all terms in the synset and all those that are only one node away from the words in the synset. When these methods were applied to a short query (the description field in Query #122), significant improvement was obtained. When they were applied to a long query containing all the material in Query #122, the improvement was not significant. In summary, Voorhees succeeded in improving the system performance significantly on short queries, but not on long queries.

9. RELATIONS IN IR EXPERIMENTS IN OTHER LANGUAGES

The Tipster project and the associated TREC Conferences pushed the information retrieval research community toward the search for retrieval methods that are valid in different languages. Although the initial emphasis was on Japanese, the community has become more receptive to work on information retrieval in many different languages. The use of relational thesauri shows promise of being a technique that can be effective in many different languages.

9.1. Japanese

Fujii and Croft (1993) report on experiments in Japanese information retrieval using the INQUERY approach that Croft devised for English. They proposed using a thesaurus as a way to handle multiple spellings of the same word, a major problem in Japanese because of the different scripts, the different versions of Kanji characters, and the large number of loan words. For example, there are at least three ways to express the concept of *running:* Hashiru-koto, which represents original Japanese usage; SouKou, a Kanji term based on a word from Chinese; and ran'ningu, a phonetic Katakana translation (1993, p. 240). They seem to be describing the relation that sometimes appears in commercial dictionaries as "variant" or "cf".

Information retrieval in both Chinese and Japanese depends on successful word segmentation, because neither language is written with word spaces. Many strings of characters are potentially ambiguous for this reason, so word segmentation and disambiguation go hand in hand. Fortunately, this complex problem is yielding to both dictionary-based approaches (Lin & Evens, 1995) and machine-learning approaches (Wu & Tseng, 1995).

9.2 Chinese

Wan, Evens, Wan, & Pao (1997) demonstrated the effectiveness of a relational thesaurus in bibliographic information retrieval in Chinese, using a corpus of 555 abstracts published by the Science and Technology center in Taiwan. This collection included all abstracts of papers in computer and information science listed by the center between June 1993 and April 1994. The experiment used an inventory of eleven relations developed by Casagrande and Hale (1967), included in table 1.

The thesaurus used in Wan's experiments was an interactive one; that is, the user could designate a word or a phrase and get back all the related words and phrases, then click on those to be added to the query. Alternatively, the user could choose a given relation R and get a screen containing R-related words only. In practice, the users only chose this option when the "all relation" option supplied more feedback than would fit on one screen. A standard recall-precision analysis showed a significant improvement in precision at all recall levels. It would be good to try Wan's approach on a much larger test corpus, like the TREC-5 test data with 28 queries and 24,988 articles from *Xinhua* and 139,801 from *People's Daily.*

9.3 Arabic

Over the last ten years there have been a series of experiments in information retrieval at the Laboratory for Arabic Language Processing at Illinois Institute of Technology. The first steps in this research project were taken by Ibrahim Al-Kharashi, who entered a number of abstracts of papers in computer science from the Proceedings of the Saudi Arabian National Computer Conferences for this purpose. Al-Kharashi did a series of experiments using words, stems, and roots as index terms and found that roots gave significantly better results than stems or words (Al-Kharashi & Evens, 1994). Hani Abu-Salem (1992) then repeated these experiments with the addition of a relational thesaurus developed by Mohammad Alkhrisat (1992).

Alkhrisat started with Fox's list of relations (table 2) but made some modifications that reflect the differences in morphology between Arabic and English. He added four relations that reflect regular derivational morphology in Arabic to the paradigmatic group.

Abu-Salem (1992) implemented the thesaurus in an interactive style. Users were presented with a list of semantically related terms and given a chance to click on those that they wished to add to the query. Abu-Salem's results using words vs. stems vs. roots as index terms without the relational thesaurus gave very similar results to those obtained by Al-Kharashi. The use of a relational thesaurus with words as index terms gave approximately the same recall as using stems or roots and much better precision. Using stems or roots as index terms along with the relational thesaurus did not give significant improvement over the results obtained using words and also using the relational thesaurus. We believe that the use of a thesaurus to enrich the queries provides the morphologically related terms that are produced by using roots or stems—but in a more discriminating way—and also provides other semantically related terms as well. This experiment should be repeated with a large collection of abstracts and with another corpus to see whether the same results are obtained.

9.4 Finnish

Kekalainen and Jarvelin (1998) describe a series of experiments using thesauri for query expansion in Finnish. Finnish is a highly agglutinative language, even more so than Japanese or Arabic. To counteract the Finnish tendency to create long compound words, they break up compounds during the matching process. During their retrieval experiments they grouped relations into four types, synonymy relations, "narrower concept" relations, and "next" relations, according to whether they tended to narrow or broaden the query. They conclude that the effectiveness of different kinds of expansion depends more on the structure of the original query than on the particular relation or relation type.

10. NEW DIRECTIONS

Renewed interest in relational thesauri has led to a number of recent experiments. Mandala, Tokunaga, and Tanoka (1999) improved retrieval results significantly by enriching queries with terms from thesauri of several different types in an experiment using the fifty queries and 528,155 documents in the TREC-7 test collection. They used a manually constructed thesaurus, based on WordNet, a co-occurrence based thesaurus, and one based on "head-modifier" relationships. Words are related in the "head-modifier" thesaurus if they appear together in one of four different syntactic relations, subject-verb, verb-object, adjective-noun, and noun-noun, and they also score above a calculated mutual information threshold. They then calculate weights for the expansion terms in these thesauri on the basis of similarity to all terms in the original query, not just one term.

Sanderson and Croft (1999) have worked on the automatic development of concept hierarchies as a vehicle for query expansion. Their approach is also useful for finding salient term pairs. Interest in using mutual information statistics (Church & Hanks, 1990) to identify collocations and collocational relations has led to several promising experiments. Jang, Myaeng, and Park (1999) have used this approach in a successful experiment in Korean-English cross language retrieval. Zhou (1999) reports on experiments in the automatic discovery of phrasal terms at Lexis/Nexis. He computed mutual information statistics for word pairs, triples, and quadruples. Phrases identified in this manner included "medical malpractice", "health care provider", "applicable standard of care".

Recently Strzalkowski, Lin, Wang, & Perez-Caballo (1999) have tried expanding queries not just with related index terms but with entire text passages. The initial results are very impressive. They expanded queries by pasting in entire sentences or paragraphs from related documents. The related documents were identified by carrying out a conventional search and picking the top ten or twenty documents retrieved in that search. Then they scanned these documents for passages that contain the concepts present in the query. Finally, they pasted these passages into the query (1999, p. 138). When the passages were chosen manually (with a human being recognizing the concepts in the query and identifying passages with the same concepts), the increase in precision was almost 40% at a given recall level. When they performed this query expansion automatically, the improvement dropped to 7%. They are now trying to figure out a way to improve the automatic process.

Gerda Ruge (1999) describes an experiment in thesaurus construction in her paper, "Combining Corpus Linguistics and Human Memory Models for Automatic Term Association". She argues that the most appropriate relations for information retrieval applications are synonymy, hyperonymy (broader terms), hyponymy (narrower terms), part-of (as in car-tank), antonymy, and compatibility (collocation relations of all kinds). Using these links, she proposes to construct a large term association network based on the spreading activation model of Collins and Loftus (1975). Smeaton (1999) discusses using WordNet links in a similar way. After all, Miller originally designed WordNet as a model of human associative memory. WordNet allows us to define semantic distance in terms of nodes traversed reaching one word from the next across WordNet links.

11. CONCLUSION

There is still no definitive inventory of relations for use in building thesauri for information retrieval. The large inventories of relations used in knowledge representation and inferencing (Weaver, 1992) do not seem to be necessary for information retrieval. On the other hand, grouping all related words together, as Salton and others did in the early experiments with thesauri is not necessarily ideal either. Both Fox (1981) and Wang, Vandendorpe, and Evens (1985) obtained better results by leaving out antonyms. It seems intuitively desirable to separate relations into groups that tend to expand the size of the pool of documents retrieved vs. those that cut down that pool vs. those that move the query 'sideways' at roughly the same level, to give the same number of responses clustered around a different target. Finer distinctions may be appropriate if we learn how to use them appropriately. More research is certainly needed in this area.

References

Abu-Salem, H. (1992). *A Microcomputer Based Arabic Information Retrieval System with Relational Thesauri (Arabic-IRS).* Doctoral dissertation, Illinois Institute of Technology.

Ahlswede, T. E. (1985). A linguistic string grammar for adjective definitions. In S. Williams (Ed.), *Humans and Machines,* 101-127. Norwood, NJ: Ablex.

Ahlswede, T. E. (1992). Issues in the design of test data for lexical disambiguation by humans and machines. *Proceedings of the Fourth Midwest Artificial Intelligence and Cognitive Science Conference,* 112-116.

Ahlswede, T. E. (1995). Word sense disambiguation by human informants. *Proceedings of the Sixth Midwest Artificial Intelligence and Cognitive Science Conference,* 83-87.

Ahlswede, T. E., & Evens, M. (1988a). Generating a relational lexicon from a machine-readable dictionary. *International Journal of Lexicography,* 1, 214-237.

Ahlswede, T. E., & Evens, M. (1988b). Parsing vs. text processing in the analysis of dictionary definitions. *25th Annual Meeting of the Association for Computational Linguistics: Proceedings of the Conference,* 217-224.

Ahlswede, T. E., & Lorand, D. (1993). Word sense disambiguation by human subjects: Computational and psycholinguistic applications. *Acquisition of Lexical Knowledge from Text: Proceedings of the ACL SIGLEX Workshop,* 1-9.

Ahlswede, T. E., Evens, M., Markowitz, J., & Rossi, K. (1986). Building a lexical database by parsing Webster's Seventh Collegiate Dictionary. *Advances in Lexicology: Proceedings of the University of Waterloo Centre for the New Oxford English Dictionary,* 1986, 65-78.

Ahlswede, T. E., Anderson, J., Evens, M., Li, S. M., Neises, J., Pin-Ngern, S., & Markowitz, J. (1988). Automatic construction of a phrasal thesaurus for an information retrieval system. *Proceedings of RIAO88,* 597-608.

Al-Kharashi, I., & Evens, M. (1994). Words, stems, and roots in an Arabic retrieval system. *Journal of the American Society for Information Science,* 45, 548-560.

Alkhrisat, M. M. (1992). Structuring the Arabic lexicon and thesaurus with lexical-semantic relations. *Midwest Artificial Intelligence and Cognitive Science Conference,* 117-121.

Amsler, R. A. (1981). A taxonomy for English nouns and verbs. *19th Annual Meeting of the Association for Computational Linguistics: Proceedings of the Conference,* 133-138.

Apresyan, Y. D., Mel'cuk, I. A., & Zholkovsky, A. K. (1970). Semantics and lexicography: Towards a new type of unilingual dictionary. In F. Kiefer (Ed.), *Studies in Syntax and Semantics,* 1-33. Dordrecht: Reidel.

Berlin, B., Breedlove, D., & Raven, P. (1966). Folk taxonomies and biological classification. *Science,* 154.3746, 273-275.

Byrd, R. J. (1989). Discovering relationships among word senses. *Proceedings of the Fifth Annual Conference of the University of Waterloo Centre for the New Oxford English Dictionary,* 67-79.

Byrd, R. J., Calzolari, N., Chodorow, M., Klavans, J., Neff, M., & Rizk, O. (1987). Tools and methods for computational lexicology. *Computational Linguistics,* 13, 219-240.

Calzolari, N. (1988). The dictionary and the thesaurus can be combined. In M. Evens (Ed.), *Relational Models of the Lexicon*, 75-96. Cambridge, UK: Cambridge University Press.

Calzolari, N., & Picchi, E. (1988). Acquisition of semantic information from an online dictionary. *Proceedings of the 12th International Conference on Computational Linguistics*, 87-92.

Calzolari, N., Baker, M., & Kruyt, Y. (Eds.). (1996). *Towards a Network of European Reference Corpora. Linguistica Computazionale* XI. Pisa: Giardini.

Casagrande, J. B., & Hale, K. L. (1967). Semantic relations in Papago folk definitions. In D. Hymes & W. E. Bittle (Eds.), *Studies in Southwestern Ethnolinguistics*, 165-196. The Hague: Mouton.

Church, K., & Hanks, P. (1990). Word association norms, mutual information, and lexicography. *Computational Linguistics,* 16, 22-29.

Collins, A., & Loftus, E. (1975). A spreading activation theory of semantic processing. *Psychological Review,* 82, 407-428.

Evens, M., & Smith, R. N. (1978). A lexicon for a computer question-answering system. *American Journal of Computational Linguistics,* 4, 1-96.

Fellbaum, C. (Ed.). (1998). *WordNet: An Electronic Lexical Database.* Cambridge, MA: MIT Press.

Fox, E. A. (1981). Lexical relations: Enhancing effectiveness of information retrieval systems. *SIGIR Forum*, 15, 5-36.

Fox, E., Nutter, J. T., Ahlswede, T., Evens, M., & Markowitz, J. (1988). Building a large thesaurus for information retrieval. *Proceedings of the Fifth Conference on Applied Natural Language Processing*, 101-108.

Fujii, H., & Croft, W. B. (1993). A comparison of indexing techniques for Japanese information retrieval. *Proceedings of the Sixteenth Annual International ACM/SIGIR Conference on Research and Development in Information Retrieval*, 237-246.

Harman, D. (Ed.) (1993-). *The Text REtrieval Conference (TREC).* [D. Harman (Ed.), TREC-1-TREC-4; D. Harman & E. Voorhees (Eds.), TREC-5-.] NIST Special Publication 500-207 (TREC-1), 500-215 (TREC-2), 500-225 (TREC-3), 500-236 (TREC-4), 500-238 (TREC-5), 500-240 (TREC-6), 500-242 (TREC-7), 500-246 (TREC-8), 500-249 (TREC-9). Gaithersburg, MD :National Institute of Standards and Technology. Available: <http://trec.nist.gov/pubs.html> [2001, October 9].

Jang, M. G., Myaeng, S. H., & Park, S. Y. (1999). Using mutual information to resolve query translation ambiguities and query term weighting. *37th Annual Meeting of the Association for Computational Linguistics: Proceedings of the Conference*, 223-229.

Jenkins, J. H. (1970). The 1952 Minnesota word association norms. In L. Postman & G. Keppel (Eds.), *Norms of Word Association*. 1-38. New York: Academic Press.

Kekalainen, J., & Jarvelin, K. (1998). The impact of query structure and query expansion on retrieval performance. *Proceedings of the 21st Annual International ACM/SIGIR Conference on Research and Development in Information Retrieval*, 130-137. Melbourne, Australia.

Lesk, M. (1987). Automatic sense disambiguation using machine-readable dictionaries: How to tell a pine cone from an ice cream cone. *Proceedings of the Fifth International Conference on Systems Documentation*, 24-26.

Lin, W. H., & Evens, M. (1995). Statistical approaches to Chinese word segmentation. *Proceedings of Midwest Artificial Intelligence and Cognitive Science*, 83-87.

Mandala, R., Tokunaga, T., & Tanoka, H. (1999). Combining multiple evidence from different types of thesaurus for query expansion. *Proceedings of the 22nd Annual International ACM/SIGIR Conference on Research and Development in Information Retrieval,* 191-197.

Mel'cuk, I. A., & Zholkovsky, A. K. (1988). The Explanatory-Combinatorial Dictionary. In M. Evens (Ed.), *Relational Models of the Lexicon,* 41-74. Cambridge, UK: Cambridge University Press.

Miller, G. A. (1998). Nouns in WordNet. In C. Fellbaum (Ed.), *WordNet: An Electronic Lexical Database,* 23-46. Cambridge, MA: MIT Press.

Miller, G., Beckwith, R., Fellbaum, C., Gross, D., & Miller, K. (1991). WordNet. *International Journal of Lexicography,* 4, 1-75.

Minker, J., Wilson, G. A., & Zimmerman, B. H. (1972). An evaluation of query expansion by the addition of clustered terms for a document retrieval system. *Information Storage and Retrieval,* 8, 329-348.

Norman, D. (1976). *Memory and Attention.* New York: Wiley.

Nutter, J. T., Fox, E., & Evens, M. (1990). Building a lexicon from machine-readable dictionaries for improved information retrieval. *Literary and Linguistic Computing,* 5, 129-138.

Pentheroudakis, J., & Vanderwende, L. (1993). Automatically identifying morphological relations in machine-readable dictionaries. *Making Sense of Words: Proceedings of the Ninth Annual Conference of the UW Center for the New OED and Text Research,* 75-88.

Riegel, K. (1970). The language acquisition process: A reinterpretation of selected research findings. In L. R. Goulet & P. B. Baltis (Eds.), *Lifespan Developmental Psychology,* 357-359. New York; Academic Press.

Ruge, G. (1999). Combining corpus linguistics and human memory models for automatic term association. In T. Strzalkowski (Ed.), *Natural Language Information Retrieval,* 75-98. Dordrecht: Kluwer.

Ruimy, N., Corazzari, O., Gola, E., Spanu, A., Calzolari, N., & Zampolli, A. (1998). European LE-PAROLE Project: The Italian Syntactic Lexicon. *Proceedings of the First International Conference on Language Resources and Evaluation,* 241-248.

Russell, W., & Jenkins, J. R. (1954). *The Complete Minnesota Norms for Responses to 100 Words on the Kent-Rosanoff Word Association Test.* (Studies in the Role of Language in Behavior Technical Report No. 11.) Minneapolis: University of Minnesota, Department of Psychology.

Salton, G. (1968). *Automatic Information Organization and Retrieval.* New York: McGraw-Hill.

Salton, G. (1973). Comment on "An evaluation of query expansion by the addition of clustered terms for a document retrieval system." *Computing Reviews,* 14, 232.

Sanderson, M., & Croft, B. (1999). Deriving concept hierarchies from text. *Proceedings of the 22nd Annual International ACM/SIGIR Conference on Research and Development in Information Retrieval,* 206-213.

Sedelow, S. Y. (Ed.). (1973). *Automated Language Analysis.* Lawrence, KS: University of Kansas.

Sedelow, S. Y. (1991). Exploring the terra incognita of whole language thesauri. *Midwest Artificial Intelligence and Cognitive Science Conference,* 108-111.

Smeaton, A. (1999). Using NLP or NLP resources for information retrieval. In T. Strzalkowski (Ed.), *Natural Language Information Retrieval*, 99-112. Dordrecht: Kluwer.

Sparck Jones, K. (1971). *Automatic Keyword Classification for Information Retrieval.* London: Butterworths.

Strzalkowski, T., Lin, F., Wang, J., & Perez-Caballo, J. (1999). Evaluating natural language processing techniques in information retrieval. In T. Strzalkowski (Ed.), *Natural Language Information Retrieval,* 113-145. Dordrecht: Kluwer.

Voorhees, E. M. (1994). Query expansion using lexical-semantic relations. *Proceedings of the Seventeenth Annual International ACM/SIGIR Conference on Research and Development in Information Retrieval,* 61-69.

Voorhees, E. M. (1998). Using WordNet for text retrieval. In C. Fellbaum (Ed.), *WordNet: An Electronic Lexical Database,* 287-303. Cambridge, MA: MIT Press.

Walker, D. E. & Amsler, R. A. (1986). The use of machine-readable dictionaries in sublanguage analysis. In R. Grishman & R. Kittredge (Eds.) *Analyzing Language in Restricted Domains,* 69-83. Hillsdale, NJ: Erlbaum.

Wan, T. L., Evens, M., Wan, Y. W., & Pao, Y. Y. (1997). Experiments with automatic indexing and a relational thesaurus in a Chinese information retrieval *system. Journal of the American Society for Information Science,* 48, 1086-1096.

Wang, G. N., Evens, M., & Hier, D. (1993). Query enhancement with relational thesauri. *Proceedings of the Seventh Annual Midwest Computer Conference,* 63-68.

Wang, Y. C., Vandendorpe, J., & Evens, M. (1985). Relational thesauri in information retrieval. *Journal of the American Society for Information Science,* 36, 15-27.

Weaver, F. (Ed.). (1992). *Computers and Mathematics with Applications* (special issue on relational models), 23, 6-9.

Wu, Z., & Tseng, G. (1995). ACTS: An automatic Chinese text segmentation system for full text retrieval. *Journal of the American Society for Information Science,* 46, 83-96.

Zhou, J. (1999). Phrasal terms in real-world IR applications. In T. Strzalkowski (Ed.), *Natural Language Information Retrieval.* 215-257. Dordrecht: Kluwer.

Chapter 10

Identifying Semantic Relations in Text
for Information Retrieval and Information Extraction

Christopher Khoo
Division of Information Studies, Nanyang Technological University, Singapore

Sung Hyon Myaeng
Department of Computer Science, Chungnam National University, Korea

Abstract:
    Automatic identification of semantic relations in text is a difficult problem, but is important for many applications. It has been used for relation matching in information retrieval to retrieve documents that contain not only the concepts but also the relations between concepts specified in the user's query. It is an integral part of information extraction—extracting from natural language text, facts or pieces of information related to a particular event or topic. Other potential applications are in the construction of relational thesauri (semantic networks of related concepts) and other kinds of knowledge bases, and in natural language processing applications such as machine translation and computer comprehension of text. This chapter examines the main methods used for identifying semantic relations automatically and their application in information retrieval and information extraction.

1. INTRODUCTION

    The ability to identify semantic relations in text automatically and accurately is increasingly important in information retrieval and information extraction systems. Information retrieval involves identifying a small subset of documents in a document collection that are likely to contain relevant information in response to the user's query statement. This is performed by matching concepts (typically keywords) in the documents with concepts in the query statement. If the query contains more than one concept, then semantic relationships between the concepts may also be specified in the query. In this case, the relevant documents to be retrieved should contain all the concepts in the query as well as the correct relationships between the concepts. Traditional information retrieval has focused on matching keywords in the text and has ignored the semantic relationships expressed between the concepts. If semantic relationships can be identified accurately in text, this can improve retrieval results for some queries by eliminating the documents containing the required concepts but not the desired relationships between the concepts.

Increasingly, users will not be satisfied with retrieval systems that retrieve whole documents that the user must then read through to identify the portion containing the desired information. Users will want systems that actually extract the information they want (information extraction) and answer questions (question answering). Retrieval systems will also be expected to help users synthesize information extracted from multiple documents to provide an overview of a subject (information visualization and automated abstraction) or to identify new relationships between facts and synthesize new knowledge (text mining and knowledge discovery). Information extracted from text can also be used to build fact databases and knowledge-bases for various purposes. All these applications require accurate identification of semantic relations in text so that the concepts extracted from a text can be connected with relations to form a coherent whole.

The main approach used to identify semantic relations in text automatically is through some kind of pattern matching. This chapter discusses pattern matching techniques for relation identification and their application in information retrieval and information extraction.

2. AUTOMATIC IDENTIFICATION OF RELATIONS

Identifying semantic relations in text involves looking for certain linguistic patterns in the text that indicate the presence of a particular relation. These patterns may be linear patterns of words and syntactic categories in the text, or graphical patterns in the underlying syntactic structure (parse tree) of the text.

To identify semantic relations in text, pattern-matching is performed to identify the segments of the text or the parts of the sentence parse trees that match with each pattern. If a text segment matches with a pattern, then the text segment is identified to contain the semantic relation associated with the pattern. A pattern typically contains one or more slots or blanks, and the parts of the text that match with the slots in the pattern represent the entities related by the semantic relation.

It is difficult to attain high accuracy in identifying semantic relations in text. The accuracy rate varies widely depending on many factors – the type of semantic relation to be identified, the domain or subject area, the type of text/documents being processed, the amount of training text available, whether knowledge-based inferencing is used, the accuracy of the syntactic pre-processing of the text, etc.

This section discusses the characteristics of linear and graphical patterns, and how they are used to identify semantic relations in text.

2.1 Linear Patterns

The following is an example of a linear pattern that describes one way the cause-effect relation can be expressed in text:

[effect] *is the result of* [cause]

The tokens with square brackets represent slots to be filled by words/phrases in the text. The slots indicate which part of the sentence represents the *cause* and which part represents the *effect* in the cause-effect relation. The pattern can match with the following sentence:

The fatal accident is the result of drunken driving.

The phrase "the fatal accident" matches with the *effect* slot, and "drunken driving" matches with the *cause* slot.

A linear pattern is thus a sequence of tokens, each token representing one or more of the following:

- *A literal word* in the text, which may or may not be stemmed or converted to a root form;
- *A syntactic constituent*, which may be
  - A part-of-speech, e.g., noun, verb and preposition;
  - A type-of-phrase, e.g., noun phrase, verb phrase, prepositional phrase and clause;
  - A syntactic role, e.g., subject, direct object, indirect object, active-voice verb and passive-voice verb;
- *A semantic category*, which may be represented by a concept label from a thesaurus or conceptual hierarchy, or a label representing a meaningful unit, e.g., organization name, person name, date and amount of money;
- *A wildcard*, that can match with one or more words in the text;
- A pre-defined set of words or syntactic/semantic categories;
- A set of sub-patterns;
- A slot or blank to be filled in by words in the text.

A token can specify multiple constraints. A token can be a triple <x, y, z>, where

- x is a symbol representing a literal word, a wildcard or a slot;
- y is a symbol representing a syntactic constraint;
- z is a symbol representing a semantic constraint.

A slot in a pattern can thus have a syntactic and/or semantic constraint attached to it.

Table 1 lists some of the patterns used by Khoo, Kornfilt, Oddy and Myaeng (1998) to identify cause-effect relations in text. The patterns listed all involve the phrase *because of*; [1] represents the cause slot, and [2] represents the effect slot. Some slots in the patterns have a syntactic constraint (e.g., noun, noun phrase, etc.) specifying what type of word or phrase may fill the slot.

For each set of patterns, the patterns are tried in the order listed in the set. Once a pattern is found to match a sentence (or some part of a sentence), all the words that match the pattern (except for the words filling the slots) are flagged, and these flagged words are not permitted to match with tokens in any subsequent pattern. So, the order in which patterns are listed in the set is important. As a rule, a more "specific" pattern is listed before a more "general" pattern. A pattern is more specific than another if it contains all the tokens in the other pattern as well as additional tokens not in the other pattern.

Pattern 2 in table 1 is an example of a negative pattern. If a sentence contains the words "not because", this will match with pattern 2 and will be flagged so that they will

No.	Relation	Pattern
1	C	[1] &and because of &this [2] &end_of_clause Example: It was raining heavily and because of this the car failed to brake in time.
2	-	&NOT because Example: It was not because of the heavy rain that the car failed to brake in time.
3	C	it &aux &adv because of [1] that [2] &end_of_clause Example: It was because of the heavy rain that the car failed to brake in time.
4	C	&beginning_of_clause [2] &and_this &aux &adv because of [1] &end_of_clause Example: The car failed to brake in time and this was because of the heavy rain.
5	C	&beginning_of_clause because of [N:1] , [2] Example: John said that because of the heavy rain , the car failed to brake in time.
6	C	&beginning_of_clause [2] because of [1] &end_of_clause Example: The car failed to brake in time because of the heavy rain.

Notes:
1. The code in the second column indicates the type of semantic relation associated with the pattern. "C" refers to the cause-effect relation. "-" indicates a negative relation.
2. [1] and [2] in the patterns represent slots to be filled by the first and second constituent of the relation, the first constituent of the cause-effect relation being the *cause* and the second constituent the *effect*. The type of phrase or word that may fill a slot may also be indicated. The symbol [N:1] indicates that the slot for *cause* is to be filled by a noun phrase.
3. The symbol & followed by a label refers to a set of subpatterns (usually a set of synonymous words or phrases). For example, *&aux* refers to auxiliary verbs like *will*, *may*, and *may have been*, and *&beginning_of_clause* refers to subpatterns that indicate the beginning of a clause.

Table 1. Examples of linear patterns for identifying the cause-effect relation

not match with any of the subsequent patterns and be identified as containing a cause-effect relation.

The patterns and the pattern matching process appear simple and straightforward. Each linear pattern is equivalent to a finite-state transition network and therefore is easy to implement. However, a large number of patterns may be needed for a particular application, and so, there must be an effective method for constructing and managing the set of patterns, and controlling the interaction between related patterns.

There are also several possible variations in the kinds of patterns used, and how the pattern matching is performed:

1. A pattern may be required to match a *whole* sentence, or just a *portion* of a sentence.
2. A pattern may be allowed to match only *one* portion of a sentence or *multiple* portions of a sentence (i.e., have multiple matches).
3. A pattern may match text *within* a sentence or *across* sentences.

4. A pattern may have a maximum of one, two, or more slots

5. A pattern may have only positive constraints or may have negative constraints as well (i.e., may specify that a particular word or syntactic/semantic category not be present in the matching text segment).

6. The set of patterns constructed may be *mutually exclusive*, i.e., if a pattern matches with a text segment, then no other pattern in the set will match the text segment or an overlapping text segment. On the other hand, the patterns may be *overlapping*, i.e., a text segment can match with multiple patterns.

7. If the patterns are overlapping, the pattern matching program can *allow multiple matches* or *single matches* only. With *multiple matches*, a text segment is allowed to match multiple patterns. With *single matches*, a text segment is allowed to match with the first matching pattern only, after which all the matching words are flagged and are not allowed to match with any other pattern. If only single matches are allowed, then the patterns in the set have to be ordered and the patterns are tried in that particular order.

8. The set of patterns may contain positive patterns only, i.e., each pattern is used to identify a particular semantic relation. Or, it may contain negative patterns as well. Negative patterns indicate that if a text segment matches it, then the text segment does not contain the specified semantic relation. When the pattern matching program allows *single matches* only (see item 7), negative patterns can be used to ensure that certain text segments (those that match with the negative patterns) do not match with any of the subsequent positive patterns.

9. If a pattern can contain sub-patterns, then a pattern may be *recursive* (i.e., the sub-pattern may refer to the parent pattern), or non-recursive (i.e., the sub-pattern cannot refer to the parent pattern).

When developing a pattern-matching system for identifying semantic relations, many decisions thus have to be made.

The advantage of using linear patterns as opposed to graphical patterns is that a complete syntactic parse tree of each sentence need not be constructed. Parsing the syntactic structure of sentences is a notoriously difficult problem, and present-day parsers still do not give accurate results for complex sentences. Moreover, only a small segment of a sentence may contain the relation of interest, and parsing a long complex sentence may be largely unnecessary effort.

Linear patterns require only partial syntactic pre-processing of sentences and a small amount of semantic processing. The amount and type of pre-processing required depends, of course, on the nature of the patterns and the kind of syntactic/semantic constraints that may be used in the patterns. One or more of the following kinds of pre-processing may be needed:

1. *Sentence boundary identification*: identifying the beginning and end of sentences;

2. *Part-of-speech tagging*: identifying the grammatical category of individual words;

3. *Phrase identification and bracketing*: recognizing major phrasal units such as noun phrases, prepositional phrases, and clauses;

4. *Partial syntactic processing* to identify syntactic roles such as subject and direct object, and active and passive constructions;

5. *Semantic tagging*:   mapping certain words to the conceptual hierarchy or identifying the semantic type of a phrase (e.g., company names, place names, currencies, etc.).

2.2 Graphical Patterns

Linear patterns have the advantage of simplicity and the syntactic pre-processing of text required for the pattern matching can be accomplished fairly easily and accurately. The disadvantage is the large number of patterns that have to be constructed to cover all the superficial variations in sentence constructions.

If a good parser is available, graphical pattern matching that identifies graphical patterns in the syntactic parse trees of sentences can be used. Since different sentence constructions can map to the same parse tree, this can reduce the number of patterns that need to be constructed. Moreover, some difficult syntactic processing such as prepositional attachment (identifying whether a prepositional phrase should be linked to a particular verb or particular noun phrase) may have been resolved by the parser, so that the patterns need not have to handle them.

Graphical patterns are usually in the form of tree structures that are used to match with parts of sentence parse trees. Each node in a graphical pattern can represent a literal word, a semantic category, a slot, a wildcard or a sub-tree. Labeled arcs in the graphical pattern represent syntactic relations between the words.

Khoo, Chan, Niu, & Ang (1999) and Khoo, Chan, & Niu (2000) used graphical pattern matching to extract cause-effect information from medical abstracts. In their study, each sentence is parsed using Conexor's *Functional Dependency Grammar of English Parser* (http://www.conexor.fi/) to generate a graph representing the syntactic structure of the sentence.

Figure 1 gives an example of a sentence and the syntactic structure output by Conexor's parser, shown as a graphical diagram as well as in *linear conceptual graph* notation (Sowa, 1984). In this notation, the concept nodes, representing words in the sentence, are in square brackets and the relation nodes, representing syntactic relations, are given in parentheses.

Figure 1 also provides an example of a graphical pattern that matches with the parse tree of the example sentence, and can be used to extract cause-effect information from the sentence. Within the square brackets, a word in quotation marks represents a stemmed word that must occur in the sentence. A word not in quotation marks (e.g., the word *improve*) refers to a class of synonymous words that can occupy that node. "*" is a wildcard character that can match any term. The roles (or semantic relations) in the cause-effect situation are indicated within the square brackets after the ":" symbol.

2.3 Identifying Case Relations

There are many different types of semantic relations. These can be at different "levels of abstraction." A high level relation can be decomposed into more primitive concepts and relations.

The sentence "*John eats an apple*" can be represented in conceptual graph

Example sentence: A removable prosthesis and a fixed partial denture are used to improve a little girl's appearance and oral function.

Syntactic structure

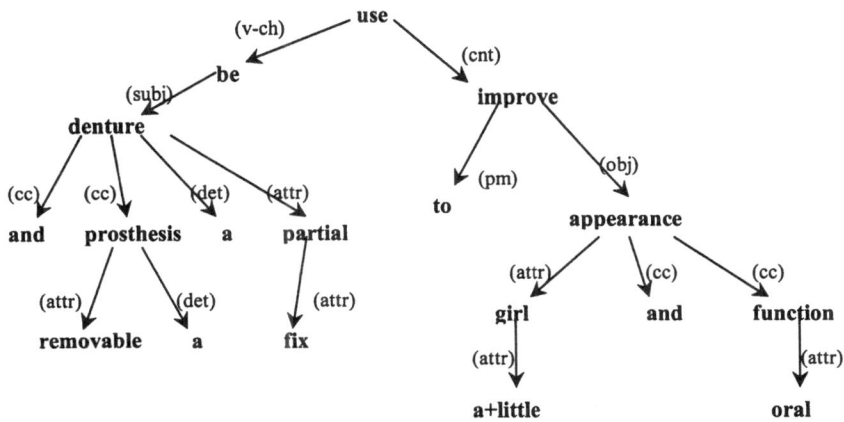

Syntactic structure in linear conceptual graph format
    [use] -
        (v-ch) → [be] → (subj) → [denture] -
            (cc) → [prosthesis] -
                (det) → [a]
                (attr) → [removable] ,
            (cc) → [and]
            (det) → [a]
            (attr) → [partial] → (attr) → [fix] ,
        (cnt) → [improve] -
            (pm) → [to]
            (obj) → [appearance] -
                (attr) → [girl] → (attr) → [a+little]
                (cc) → [and]
                (cc) → [function] → (attr) → [oral] , , , .

Relation node abbreviations	
attr:	attributive nominal
cc:	coordinating conjunction
cnt:	contingency
det:	determiner
obj:	object
pm:	preposed marker
subj:	subject
v-ch:	verb chain

The parts of the sentence that match with the graphical pattern below are indicated in bold.

Example pattern
    [*: *cause.object*] ← (subj) ← ["be"] ← (v-ch) ← ["use"] → (cnt) → [improve: *effect.polarity*] → (obj) → [*: *effect.object*]

Information extracted
    Cause.object: denture
    Effect.object: appearance
    Effect.polarity: improve

Figure. 1. Sentence structure and matching pattern in linear conceptual graph format

representation (Sowa, 1984) as

[person: "John"] → (eat) → [apple]

The diagram indicates that John has an "eating" relationship with an apple. The relation *eat* is a high-level relation that can be decomposed into the concept *eat* and the "case relations" *agent* and *patient*:

[person: "John"] ← (agent) ← [eat] → (patient) → [apple]

Case relations are low-level semantic relations that exist between the main verb of a clause and the other constituents of the clause (Fillmore, 1968; Somers, 1987). In the Case Theory, case roles (also called *thematic roles*) are assigned by the verb to the various syntactic constituents of the clause (i.e., the subject, direct object, indirect object, prepositional phrase, etc.). For example, in the sentence,

Mary bought a watch for John.

the case relations between the verb *buy* and the other constituents of the sentence are:

buy –
    → (agent) → [Mary]
    → (patient) → [watch]
    → (recipient) → [John]

Many natural language processing systems work with case relations because case relations correspond closely to the syntactic structure of sentences and are thus easier to extract from text. They are also domain-independent low-level relations that can provide a foundation for any high-level application. Identifying case relations in text serves as a useful intermediate step in the conversion of text to a semantic representation. After the case relations are identified, higher-level relations can be inferred. This section examines how case relations are identified in sentences.

The linguistic patterns used for identifying case relations are called *case frames*. One or more case frames are constructed for each sense or meaning of a verb. A case frame specifies:

1. The number of entities that the verb expects in the sentence,
2. The case roles assigned to these entities,
3. "Selectional restrictions" that specify the semantic category of the entity filling a role,
4. The syntactic position of each entity in the sentence, and
5. Whether each role is obligatory (i.e., must be filled) or optional.

Table 2 gives examples of case frames for the verbs *buy* and *decline*. The first case frame (for "buy") indicates that the subject of the verb has an *agent* role, and the entity filling this role must belong to the semantic class of human entities. The entity assigned the patient role must be a concrete object. The recipient role is assigned to the object of the preposition "for". All the roles are obligatory, indicating that if all the roles are filled and the entities filling the roles satisfy the syntactic and semantic constraints, then the case frame is instantiated, and there is a match. We can then infer that the semantic relations

Frame no.	Verb	Sense no.	Case role	Syntactic category	Semantic category	Obligatory / Optional
1	Buy	1	Agent Patient Recipient	Subject Object Prepositional phrase "for"	Human Concrete Human	Obligatory Obligatory Obligatory
2	Decline (e.g., "The stock prices declined")	1	Patient	Subject	—	Obligatory
3	Decline (e.g., "She declined the offer")	2	Agent Patient	Subject Object	Human —	Obligatory Optional
4	Decline (e.g., "She declined to go.")	3	Agent Activity	Subject Infinitive	Human —	Obligatory Obligatory

Table 2. Example case frames

between the main verb and the various constituents of the clause are the case relations specified in the case frame.

If there is no match, then the system has to find another case frame to match the clause. If more than one case frame matches a clause, then the system has to select the frame that best matches it. Myaeng, Khoo, and Li (1994) used the following rules to select the best frame:

- The frame with the smallest number of obligatory rules left unfilled;
- The frame with the highest number of roles filled (i.e., the case frame which accounts for the highest number of constituents in the sentence);
- The case frame for the most commonly used sense of the verb.

Whereas a complete case frame specifies all the case roles assigned by the verb, researchers have worked with linear patterns that can be characterized as partial case frames that seek to identify just one case relation at a time. The following example patterns are derived from Cowie & Lehnert (1996):

- <"robbed", active-voice verb> <slot: robbery.victim, noun phrase>
  The noun phrase following "robbed" (active voice) refers to the robbery victim.
- <slot: kidnapping.victim, noun phrase> <"disappeared", active-voice verb>
  The noun phrase preceding "disappeared" (active voice) refers to the victim of a kidnaping.
- <"traveling", active-voice verb> <"in"> <slot: attack.target, noun phrase>
  The noun phrase following the preposition "in" after "travelling" (active voice) is the target of an attack (e.g., "The group was travelling in a four-wheel-drive vehicle.")
- <slot: attack.instrument, subject> <"hurled", passive-voice verb>

The noun phrase preceding "hurled" (passive voice) is the instrument of an attack (e.g., "Dynamite sticks were hurled from a car.").

Each of the patterns has only one slot. The generic case relation labels (*agent, patient*, etc.) had been replaced in the example patterns with higher-level domain-specific role labels that were more meaningful in the context of the application—extracting information on terrorist acts. The *patient* role for the verb *rob* was re-labeled *robbery.victim*, and the patient role for the verb *disappeared* was re-labelled *kidnapping.victim*.

3. PATTERN CONSTRUCTION

3.1 Approaches to Pattern Construction

Construction of patterns can be done manually or automatically by analyzing sample relevant text and the associated answer key indicating the semantic relation present in the text and the concepts in the text connected by the relation. The answer keys are typically constructed by human analysts that have to be trained for the task.

Pattern construction thus entails constructing patterns that would extract the same information from the text as the analysts did. The patterns constructed have to be general enough to extract correct information from unseen text segments that are "similar" to but not exactly the same as those in the training text. At the same time, the patterns should not be too general as to extract information from non-relevant text fragments or extract incorrect information from relevant text fragments. The set of patterns should be parsimonious: There should be enough patterns to give good results but not more patterns than necessary, as this will increase processing time.

Two approaches can be used in the pattern construction:

1. *General-to-specific approach*: General patterns are first constructed which are gradually specialized to reduce errors.
2. *Specific-to-general approach*: Specific patterns are first constructed which are gradually generalized to cover more situations.

The *general-to-specific approach* constructs a general pattern (pattern with the least number of constraints) from the training text. More and more constraints are added to this general pattern as negative examples are encountered. Negative examples refer to text segments that would match with the pattern, but yield incorrect information. Constraints are added to the general pattern so that it will either extract the correct information from a negative example or would not match the example at all. At the same time, care is taken to ensure that the pattern will continue to extract the correct information from previously encountered positive examples.

To specialize a general pattern, the pattern can be modified in the following ways:

1. Add a syntactic or semantic constraint to one or more tokens in the pattern.
2. Replace a semantic constraint with a more specific semantic category (e.g., change "animate" to "human").
3. Add one or more tokens to the pattern.

The general-to-specific approach is thus an optimistic approach that favors recall, i.e., it tries not to miss any relevant information at the cost of precision. It may extract spurious information or extract more information than is correct.

The *specific-to-general approach* is a conservative approach in which a pattern is constructed that will match only text fragments that very closely resemble the training text. As new example text fragments are found that indicate that the pattern is not general enough, then the constraints are gradually relaxed or removed to accommodate the new examples. These new examples can be characterized as positive examples. They do not match with the current set of patterns, which have to be modified to "cover" them and extract correction information from them. This approach favors precision at the cost of recall.

To generalize a pattern to cover more example text segments, the pattern can be modified in the following ways:

1. Remove a syntactic or semantic constraint from a token.
2. Convert a literal token (i.e., a token that has to match an exact word in the text) to a syntactic or semantic category.
3. Replace a semantic constraint with a more general semantic category (e.g., change "human" to "animate").
4. Add an alternative syntactic or semantic constraint, i.e., replace a single constraint with a disjunction of two or more alternative constraints so that only one of the constraints needs to be met. For example, a token could specify a noun phrase OR a prepositional phrase.

Another way to generalize a set of patterns is to construct a new pattern to cover the new example text segment.

In practice, some combination of these two approaches are used in order to balance recall and precision. Based on one or more similar text segments and the associated answer keys, one or more patterns are constructed that represent an estimate of the optimal level of generality and specificity. As positive and negative example texts are encountered, the patterns are generalized and specialized to improve coverage while keeping the error rate low. If the pattern construction is performed by human analysts, the analysts will use their linguistic knowledge and judgment to construct patterns at a good level of generality and specificity, perhaps with the help of some guidelines for consistency.

However, some researchers have successfully used automatic or machine-aided means of pattern construction. Automatic pattern construction probably requires a larger set of training examples with answer keys created by human analysts. Creating answer keys is probably faster and requires less training than constructing a good set of patterns. Moreover, recall and precision probabilities can be used to guide the automatic pattern construction to get optimal results.

To perform automatic pattern construction, the system needs well-defined heuristics for constructing the initial patterns and for generalizing and specializing the patterns based on positive and negative examples. We have listed above several ways of generalizing and specializing a set of patterns. The system will need heuristics for selecting which generalization/specialization method to use in which situation and the order in which the methods are tried. Typically, a variation of the inductive learning algorithm described by Michalski (1983) and Mitchell (1982) is used. In the rest of the section, we examine the heuristics used by three systems, AutoSlog, PALKA, and CRYSTAL, for automatic pattern construction.

3.2 Automatic Pattern Construction in Three Systems

The AutoSlog system (Riloff, 1993, 1996) uses partial case frames as linear patterns. Each pattern has only one slot and usually includes a verb and a noun phrase (a subject or direct object). A set of pattern templates define the linear patterns that the system will construct. Each pattern is thus an instantiation of a pattern template. Table 3 lists the pattern templates used in AutoSlog and an example pattern for each template.

Before pattern construction, the training corpus has to be preprocessed by identifying clause boundaries and the major syntactic constituents—subject, verb, direct object, noun phrases and prepositional phrases. Relevant text segments that contain the semantic relations of interest are identified, and answer keys are constructed to indicate which noun phrase should be extracted and the semantic role it has. If the domain of interest is terrorist activities, the semantic roles would include *perpetrators*, *targets*, *victims*, etc.

During pattern construction, pattern matching is used to match the pattern templates with the training text segments. If a pattern template matches with a relevant text segment, then a pattern is constructed by replacing the tokens in the template with the words in text. If a token in the template indicates a slot, this token is allowed to match a noun phrase in the text only if the noun phrase appears in the answer key (i.e., a human analyst has indicated that this is the information to be extracted). A slot token is placed in the pattern being constructed, and the semantic role for the slot is taken from the answer key. The constructed pattern is thus a specialization of the pattern template. Finally, a human analyst manually inspects each pattern and decides which ones should be accepted or rejected.

Pattern Template	Example Pattern
<subj: slot> <passive-verb>	[victim] was murdered
<subj: slot > <active-verb>	[perpetrator] bombed
<subj: slot > <verb> <infinitive-phrase>	[perpetrator] attempted to kill
<subj: slot > <auxiliary-verb> <noun>	[victim] was victim
<passive-verb> <direct-obj: slot>	killed [victim]
<active-verb> <direct-obj: slot>	bombed [target]
<infinitive-phrase> <direct-obj: slot> <verb>	to kill [victim]
<infinitive-phrase> <direct-obj: slot>	tried to attack [target]
<gerund> <direct-obj: slot>	killing [victim]
<noun> <auxiliary-verb> <direct-obj: slot> <noun>	fatality was [victim]
<preposition> <noun-phrase: slot> <active-verb>	bomb against [target]
<preposition> <noun-phrase: slot>	killed with [instrument]
<passive-verb> <preposition> <noun-phrase: slot>	was aimed at [target]

Table 3. Pattern templates and example of instantiated patterns in AutoSlog

AutoSlog does not adjust the patterns by generalizing or specializing them once they are constructed. We now look at two systems, PALKA and CRYSTAL, that do adjust the patterns after the initial construction.

In the PALKA system (Kim, 1996; Kim & Moldovan, 1995), the patterns involve the whole clause. Sentences in the training text are first converted to simple clauses. The clauses containing a semantic relation of interest are processed one at a time. If the set of patterns already constructed do not match a clause, then a new pattern is constructed for the clause. This initial pattern covers the main verb, the subject, the object, and the words to be extracted (i.e., the slot). Each of these constituents in the clause is represented by a token in the pattern. Each token is assigned a semantic category from a conceptual hierarchy.

Generalizations and specializations of the patterns are applied only to the semantic constraints. When two similar patterns are generated, their semantic constraints are generalized by locating a broader concept or ancestor in the conceptual hierarchy that is common to both semantic categories. If it is not possible to find a common broader concept, a disjunction of concepts (i.e., a set of alternative concepts) is specified as the semantic constraint. When a negative example text is found indicating that a pattern is over-generalized, the semantic constraint is specialized by finding a disjunction of concepts that covers all the positive examples but excludes the negative example.

The CRYSTAL system (Soderland, Fisher, Aseltine, & Lehnert, 1996) uses a similar approach, but is more complex. The text is also pre-processed to identify the major syntactic constituents, subject, verb phrase, direct and indirect object, and prepositional phrases. For each word in the text, its semantic class in a conceptual hierarchy is also identified.

Initially, a very specific pattern is constructed for every sentence in the training text. The sentences are not simplified into simple clauses. The constraints in the initial patterns are gradually relaxed to increase their coverage and to merge similar patterns. CRYSTAL identifies possible generalizations by locating pairs of highly similar patterns. This similarity is measured by counting the number of relaxations required to unify the two patterns. A new pattern is created with constraints relaxed just enough to unify the two patterns—dropping constraints that the two do not share and finding a common ancestor of their semantic constraints. The new pattern is tested against the training corpus to make sure it does not extract information not specified in the answer keys. If the new pattern is valid, all the patterns covered by the new pattern are deleted. This generalization proceeds until a pattern that exceeds a specified error threshold is generated.

The next section surveys how the automatic identification of relations is used in information retrieval and information extraction.

4. APPLICATIONS

4.1 Relation Matching in Information Retrieval

As mentioned earlier, information retrieval is typically performed by matching concepts (usually keywords) in the query with concepts in the documents. A document will be retrieved if it contains the query keywords even if the keywords do not have the desired relation between them as expressed in the query. In relation matching, both the concepts and the *relations* between the concepts expressed in the query are matched with concepts and relations in documents. It appears plausible that relation matching should improve information retrieval effectiveness because it provides additional evidence for predicting the document's relevance.

Early researchers in information retrieval have attempted to perform relation matching using manually identified relations. In some indexing and abstracting databases, documents are indexed using manually assigned indexing terms that are "precoordinated", i.e., some human indexer has indicated that there is a relationship between the concepts in the document. The type of relation is usually not specified in precoordinated indexing but is implied by the context. This is the case with faceted classification schemes like Precis (Austin, 1984) and Ranganathan's Colon classification (Kishore, 1986; Ranganathan, 1965). Precoordinate indexing allows the user to specify that there is some kind of relation between two terms when searching the database.

Farradane (1967) advocated the use of explicitly specified relations in the indexing system. He pointed out that implied relations in precoordinate indexing are unambiguous only in a narrow domain. In a heterogenous database covering a wide subject area, the relation between the precoordinated terms may be ambiguous. Two indexing systems that make explicit use of relations are Farradane's (1950, 1952 ,1967) relational classification system and the SYNTOL model (Gardin, 1965; Levy, 1967). It is not known whether the use of explicit relations in indexing really improves retrieval effectiveness compared with using individual index terms or using implicit relations.

Researchers have attempted to identify syntactic and semantic relations in text automatically for improving information retrieval effectiveness. Some studies have found a small improvement when syntactic relations are taken into account in the retrieval process (Croft, 1986; Croft, Turtle, & Lewis, 1991; Dillon & Gray, 1983; Smeaton & van Rijsbergen, 1988). *Syntactic relations* refer to relations that are derived from the syntactic structure of the sentence. However, the retrieval performance from syntactic relation matching appears to be no better than and often worse than the performance obtainable using index phrases generated using statistical methods based on word proximity or co-occurrence in text, such as those described in Salton, Yang, & Yu (1975) and Fagan (1989).

Semantic relations, if identified accurately, can yield better retrieval results than syntactic relations. This is because a semantic relation can be expressed using different syntactic structures. A semantic relation is partly but not entirely determined by the syntactic structure of the sentence. Syntactic relation matching can fail to find a match between similar semantic relations if the relations are expressed using different syntactic structures.

Most of the information retrieval systems that automatically identify semantic relations are information extraction systems, described in the next section. However, some researchers have studied particular types of semantic relations. Lu (1990) investigated the use of *case relation* matching using a small test database of abstracts. Using a tree-matching method for matching relations, he obtained worse results than from vector-based keyword matching.

Liu (1997) has investigated *partial relation matching*. Instead of trying to match the whole concept-relation-concept triple (i.e., both concepts as well as the relation between them), he sought to match individual concepts together with the semantic role that the concept has in the sentence. In other words, instead of trying to find matches for "term1 → (relation) → term2", his system sought to find matches for "term1 → (relation)" and "(relation) → term2" separately. Liu focused on case relations and was able to obtain positive results only for long queries (abstracts that were used as queries).

Khoo, Myaeng, and Oddy (2000) carried out an in-depth study of just one relation—the cause-effect relation. Causal relation matching did not yield better retrieval results than word proximity matching. However, as with the study by Liu (1997), partial relation matching where one concept in the relation is replaced with a wildcard was found to be helpful. Thus, if a query contains the relation "smoking causes cancer", then better retrieval results are likely to be obtained if we look for "smoking → (cause) → *" and "* → (cause) → cancer" (where "*" is a wildcard), rather than "smoking → (cause) → cancer". Substituting a wildcard for one constituent of the relation is effectively the same as assigning semantic roles to terms, as used by Liu (1997) and Marega and Pazienza (1994).

This suggests that relation matching in information retrieval may be useful for queries in which one concept in the relation is not specified, but is expected to be supplied by the relevant document, as in this example query:

I want documents that describe the consequences of the Gulf War.

This query can be represented as follows:

Gulf War → (cause) → *

4.2 Relations in Information Extraction

Whereas information retrieval seeks to identify documents that are likely to contain relevant information, information extraction goes further to identify the passage(s) in the document containing the desired information, extract the pieces of information, and relate them to one another by filling a structured template or a database record. The goal of information extraction is to find and relate relevant information while ignoring extraneous and non-relevant information (Cardie, 1997; Cowie & Lehnert, 1996; Gaizauskas & Wilks, 1998).

An information retrieval system processes ad hoc queries, i.e., the user enters any query into the system and expects to get the documents almost instantly. On the other hand, an information extraction system requires extensive training on the topic of interest before it can extract relevant information on that topic. An extensive knowledge-base has to be custom-developed for the topic. The knowledge-base will include a set of linguistic patterns for extracting relational information (as described in Sections 2 and 3) and possibly a conceptual hierarchy for the domain. Topics used for information extraction topics have been characterized as "frozen queries" (Gaizauskas & Wilks, 1998). They persist over a period of time and must be of interest to a community of users, since substantial effort is required to train the system on the topic.

Research in information extraction has been influenced tremendously by the series of Message Understanding Conferences (MUC), organized by the U.S. Advanced Research Projects Agency (ARPA). Participants of the conferences develop systems to perform common information extraction tasks, defined by the conference organizers.

For each task, a template is specified that indicates the slots to be filled in. The type of information to be extracted to fill each slot is specified, and the set of slots defines the various entities, aspects, and roles relevant to the topic of interest. An example template used in the MUC-3 Conference for terrorist activities is given in table 4. More recent MUC conferences have used more complex object-oriented templates in which a slot-fill can be the exact words extracted from the text ("string fill"), one of a given set of alternatives ("set fill"), words extracted from the text and transformed into a canonical form, e.g., dates and monetary amounts ("normalized entries"), and pointers to other objects ("references").

Text
Bogota, 30 Aug 89 (Inravision Television Cadena 2)—Last night's terrorist target was the Antioquia Liqueur Plant. . . . The watchmen on duty reported that at 2030 they saw a man and a woman leaving a small suitcase near the fence that surrounds the plant. The watchmen exchanged fire with the terrorists who fled leaving behind the explosive material that also included dynamite and grenade rocket launchers. Metropolitan Police personnel specializing in explosives defused the rockets. Some 100 people were working inside the plant.

Filled template

MESSAGE ID:	TST1-MUC3-0004
TEMPLATE ID:	1
DATE OF INCIDENT:	29 Aug 89
TYPE OF INCIDENT:	Attempted bombing
CATEGORY OF INCIDENT:	Terrorist act
PERPETRATOR: ID OF INDIV(S):	"man", "woman"
PHYSICAL TARGET: ID(S):	"Antioquia Liqueur Plant"
PHYSICAL TARGET: TOTAL NUM:	1
PHYSICAL TARGET: TYPE(S):	Commercial
HUMAN TARGET(S): ID(S):	"people"
HUMAN TARGET: TOTAL NUM:	Plural
HUMAN TARGET: TYPE(S)	Civilian
LOCATION OF INCIDENT:	Colombia: Antioquia
EFFECT ON PHYSICAL TARGET(S):	No damage: "Antioquia Liqueur Plant"
EFFECT ON HUMAN TARGET(S):	No injury or death: "people"

Table 4. Example of a filled template for terrorist acts
(Source: MUC-3, 1991, Appendix D and E)

The process of information extraction can be seen as a form of automatic identification of relations between entities, events, and situations. Whereas in Sections 2 and 3, we focused on simple relations between two entities/events/situations or a set of case relations involving a verb and other constituents of a clause, information extraction seeks to identify a web of relationships relating to a particular type of event, situation, or topic. These web of relations are encoded in the structure of the template, showing how the various pieces of information are related to one another.

Although information extraction seeks to identify many kinds of relations that are relevant to the topic of interest, the approach used for identifying relations is similar to that described earlier in the chapter. Typically, each *concept* --→ *relation* → *concept* triple is identified one at a time using pattern matching. The *concept-relation-concept* triples are merged or linked together into a bigger structure using discourse processing, co-reference resolution, or knowledge-based inferencing or are assumed to be related simply because they occur in close proximity in the document.

Information that has been extracted can be used for several purposes: for populating a database of facts about entities (e.g., companies) or events, for automatic summarization, for information mining, and for acquiring knowledge to use in a knowledge-based system. Information extraction systems have been developed for a wide range of tasks, including:

- Extracting information about international terrorist events from newswire stories (MUC-3, 1991);
- Extracting information about corporate mergers, acquisitions, and joint ventures from newspaper text (Rau, 1987; Rau, Jacobs & Zernik, 1989; MUC-5, 1994);
- Extracting information about developments in semiconductor fabrication techniques (MUC-5, 1993);
- Extracting the details of patent claims from patent documents to store in a relational database (Nishida & Takamatsu, 1982);
- Extracting information from X-ray reports for the purpose of indexing and retrieving the reports and pictures (Berrut, 1990);
- Extracting information from patient medical records and discharge summaries (Sager, 1981; Sager, Lyman, Tick, Nhàn, & Bucknall, 1994; Soderland, Aronow, Fisher, Aseltine, & Lehnert, 1995); and
- Extracting information from life insurance applications (Glasgow, Mandell, Binney, Ghemri, & Fisher, 1998).

How accurate are information extraction systems? This depends on the difficulty and complexity of the information extraction task. For fairly complex information extraction tasks, one can expect an accuracy of less than 60%, measured using an F-measure (van Rijsbergen, 1979) that weights recall and precision equally (MUC-6, 1995; MUC-7, 2000). The precision is usually below 70% and the recall usually below 60% for the best systems. Cowie and Lehnert (1996) suggest that 90% precision will be necessary for information extraction systems to satisfy users.

5. CONCLUSION

This chapter has examined how semantic relations in text are identified and extracted automatically using pattern matching with a set of linguistic patterns. We have analyzed the different types of patterns and pattern matching that can be used, as well as the different ways of constructing the patterns automatically using inductive learning. Application to information retrieval and information extraction was also surveyed.

Information retrieval systems can process ad hoc queries in real time, but perform retrieval only at the document level. Information extraction systems can extract specific information but only after extensive training. The big challenge is to combine information retrieval and information extraction technologies in a way that takes advantage of their individual strengths so that information extraction and question answering can be performed in real time for ad hoc queries.

Automatic extraction of semantic relations from text is an exciting and increasingly important area for research and development because of the tremendous amount of textual information and textual databases accessible on the World Wide Web. Information extraction for specific topics and applications has been shown to be feasible in the series of MUC conferences. Information extraction can be applied to the Web for building knowledge bases for various kinds of applications.

In particular, the World Wide Web presents tremendous opportunities for information mining and knowledge discovery. Techniques for identifying semantic relationships and extracting information can be used for automatic summarization of Web documents. Information extracted from Web documents can be chained or connected to give a conceptual map of the information available on a particular topic. The conceptual map can show how information in one document is related to information in another document and how the documents are related in content. This conceptual map can support creativity by suggesting hypotheses for investigation and indicating gaps in knowledge.

References

Austin, D. (1984). *PRECIS: A Manual of Concept Analysis and Subject Indexing* (2nd ed.). London: British Library, Bibliographic Services Division.

Berrut, C. (1990). Indexing medical reports: The RIME approach. *Information Processing & Management,* 26, 93-109.

Cardie, C. (1997). Empirical methods in information extraction. *AI Magazine,* 18(4), 65-79.

Cowie, J., & Lehnert, W. (1996). Information extraction. *Communications of the ACM,* 39(1), 80-91.

Croft, W. B. (1986). Boolean queries and term dependencies in probabilistic retrieval models. *Journal of the American Society for Information Science,* 37, 71-77.

Croft, W. B., Turtle, H. R., & Lewis, D. (1991). The use of phrases and structured queries in information retrieval. *Proceedings of the Fourteenth Annual International ACM/SIGIR Conference on Research and Development in Information Retrieval,* 32-45.

Dillon, M., & Gray, A. S. (1983). FASIT: A fully automatic syntactically based indexing system. *Journal of the American Society for Information Science,* 34, 99-108.

Fagan, J. L. (1989). The effectiveness of a nonsyntactic approach to automatic phrase indexing for document retrieval. *Journal of the American Society for Information Science,* 40, 115-132

Farradane, J. E. L. (1950). A scientific theory of classification and indexing and its practical applications. *Journal of Documentation,* 6, 83-99.

Farradane, J. E. L. (1952). A scientific theory of classification and indexing: Further considerations. *Journal of Documentation,* 8, 73-92.

Farradane, J. E. L. (1967). Concept organization for information retrieval. *Information Storage and Retrieval,* 3, 297-314.

Fillmore, C. J. (1968). The case for case. In E. Bach & R. T. Harms (Eds.), *Universals in Linguistic Theory,* 1-88. New York: Holt, Rinehart and Winston.

Gaizauskas, R., & Wilks, Y. (1998). Information extraction beyond document retrieval. *Journal of Documentation,* 54, 70-105.

Gardin, J.-C. (1965). *SYNTOL.* New Brunswick, NJ: Graduate School of Library Service, Rutgers University.

Glasgow, B., Mandell, A., Binney, D., Ghemri, L., & Fisher, D. (1998). MITA: An information-extraction approach to the analysis of free-form text in life insurance applications. *AI Magazine*, 19(1), 59-71

Khoo, C., Chan, S., Niu, Y., & Ang, A. (1999). A method for extracting causal knowledge from textual databases. *Singapore Journal of Library & Information Management*, 28, 48-63.

Khoo, C., Chan, S., & Niu, Y. (2000). Extracting causal knowledge from a medical database using graphical patterns. *38th Annual Meeting of the Association for Computational Linguistics: Proceedings of the Conference*, 336-343.

Khoo, C., Kornfilt, J., Oddy, R., & Myaeng, S. H. (1998). Automatic extraction of cause-effect information from newspaper text without knowledge-based inferencing. *Literary & Linguistic Computing*, 13, 177-186.

Khoo, C., Myaeng, S. H., & Oddy, R. (2000). Using cause-effect relations in text to improve information retrieval precision. *Information Processing & Management*, 37, 119-145.

Kim, J.-T. (1996). Automatic phrasal pattern acquisition for information extraction from natural language texts. *Journal of KISS (B), Software and Applications*, 23, 95-105.

Kim, J.-T., & Moldovan, D. I. (1995). Acquisition of linguistic patterns for knowledge-based information extraction. *IEEE Transactions on Knowledge and Data Engineering*, 7, 713-724.

Kishore, J. (1986). *Colon Classification: Enumerated & Expanded Schedules Along with Theoretical Formulations*. New Delhi: Ess Publications.

Levy, F. (1967). On the relative nature of relational factors in classifications. *Information Storage & Retrieval*, 3, 315-329.

Liddy, E. D., & Myaeng, S. H. (1993). DR-LINK's linguistic-conceptual approach to document detection. In *The First Text REtrieval Conference (TREC-1)* (NIST Special Publication 500-207), 1-20. Gaithersburg, MD: National Institute of Standards and Technology. Available: <http://trec.nist.gov/pubs.html> [2001, October 9].

Harman, D. (Ed.) (1993-). *The Text REtrieval Conference (TREC)*. [D. Harman (Ed.), TREC-1-TREC-4; D. Harman & E. Voorhees (Eds.), TREC-5-.] NIST Special Publication 500-207 (TREC-1), 500-215 (TREC-2), 500-225 (TREC-3), 500-236 (TREC-4), 500-238 (TREC-5), 500-240 (TREC-6), 500-242 (TREC-7), 500-246 (TREC-8), 500-249 (TREC-9). Gaithersburg, MD :National Institute of Standards and Technology. Available: <http://trec.nist.gov/pubs.html> [2001, October 9].

Liu, G. Z. (1997). Semantic vector space model: Implementation and evaluation. *Journal of the American Society for Information Science*, 48, 395-417.

Lu, X. (1990). An application of case relations to document retrieval. Doctoral dissertation, University of Western Ontario.

Marega, R., & Pazienza, M. T. (1994). CoDHIR: An information retrieval system based on semantic document representation. *Journal of Information Science*, 20, 399-412.

Michalski, R. (1983). A theory and methodology of inductive learning. *Artificial Intelligence*, 20, 111-161.

Mitchell, T. (1982). Generalization as search. *Artificial Intelligence*, 18, 203-226.

MUC-3. (1991). *Third Message Understanding Conference (MUC-3)*. San Mateo, CA: Morgan Kaufmann.

MUC-4. (1992). *Fourth Message Understanding Conference (MUC-3)*. San Mateo, CA: Morgan Kaufmann.

MUC-5. (1993*). Fifth Message Understanding Conference (MUC-5)*. San Francisco: Morgan Kaufmann.

MUC-6. (1995). *Sixth Message Understanding Conference (MUC-6)*. San Francisco: Morgan Kaufmann.

MUC-7. (2000). *Message Understanding Conference Proceedings (MUC-7)* [Online]. Available: http://www.muc.saic.com/proceedings/muc_7_toc.html.

Myaeng, S. H., & Liddy, E. D. (1993). Information retrieval with semantic representation of texts. *Proceedings of the 2nd Annual Symposium on Document Analysis and Information Retrieval*, 201-215.

Myaeng, S. H., Khoo, C., & Li, M. (1994). Linguistic processing of text for a large-scale conceptual information retrieval system. *Conceptual Structures: Current Practices: Second International Conference on Conceptual Structures, ICCS '94*, 69-83.

Nishida, F., & Takamatsu, S. (1982). Structured-information extraction from patent-claim sentences. *Information Processing & Management*, 18, 1-13.

Ranganathan, S. R. (1965). *The Colon Classification*. New Brunswick, N.J.: Graduate School of Library Service, Rutgers University.

Rau, L. (1987). Knowledge organization and access in a conceptual information system. *Information Processing & Management*, 23, 269-283.

Rau, L. F., Jacobs, P. S., & Zernik, U. (1989). Information extraction and text summarization using linguistic knowledge acquisition. *Information Processing & Management*, 25, 419-428.

Riloff, E. (1993). Automatically constructing a dictionary for information extraction tasks. *Proceedings of the Eleventh National Conference on Artificial Intelligence*, 811-816.

Riloff, E. (1996). An empirical study of automated dictionary construction for information extraction in three domains. *Artifical Intelligence*, 85, 101-134

Sager, N. (1981). *Natural Language Information Processing: A Computer Grammar of English and its Applications*. Reading, MA: Addison-Wesley.

Sager, N., Lyman, M., Tick, L. J., Nhàn, N. T., Bucknall, C. E. (1994). Natural language processing of asthma discharge summaries for the monitoring of patient care. *Proceedings of Seventeenth Annual Symposium on Computer Applications in Medical Care*, 265-268.

Salton, G., Yang, C. S., & Yu, C. T. (1975). A theory of term importance in automatic text analysis. *Journal of the American Society for Information Science*, 26, 33-44.

Soderland, S., Aronow, D., Fisher, D., Aseltine, J., & Lehnert, W. (1995). *Machine Learning of Text-analysis Rules for Clinical Records* (Technical Report, TE-39). Amherst, MA: University of Massachusetts, Dept. of Computer Science.

Soderland, S., Fisher, D., Aseltine, J. & Lehnert, W. (1996). Issues in inductive learning of domain-specific text extraction rules. In S. Wermter, E. Riloff, & G. Scheler (Eds.), *Connectionist, Statistical and Symbolic Approaches to Learning for Natural Language Processing*, 290-301. Berlin: Springer Verlag.

Smeaton, A. F., & van Rijsbergen, C. J. (1988). Experiments on incorporating syntactic processing of user queries into a document retrieval strategy. *Proceedings of the Eleventh Annual International ACM/SIGIR Conference on Research and Development in Information Retrieval*, 31-51.

Somers, H. L. (1987). *Valency and Case in Computational Linguistics*. Edinburgh: Edinburgh University Press.

Sowa, J. F. (1984). *Conceptual Structures: Information Processing in Mind And Machine*. Reading, MA: Addison-Wesley.

Van Rijsbergen, C. J. (1979). *Information Retrieval*. London: Butterworths.

Chapter 11

A Conceptual Framework for the Biomedical Domain

Alexa T. McCray & Olivier Bodenreider
National Library of Medicine, Bethesda, Md, USA

Abstract:
Specialized domains often come with an extensive terminology, suitable for storing and exchanging information, but not necessarily for knowledge processing. Knowledge structures such as semantic networks, or ontologies, are required to explore the semantics of a domain. The UMLS project at the National Library of Medicine is a research effort to develop knowledge-based resources for the biomedical domain. The Metathesaurus is a large body of knowledge that defines and inter-relates 730,000 biomedical concepts, and the Semantic Network defines the semantic principles that apply to this domain. This chapter presents these two knowledge sources and illustrates through a research study how they can collaborate to further structure the domain. The limits of the approach are discussed.

1. INTRODUCTION

The Unified Medical Language System® (UMLS®) project at the U.S. National Library of Medicine (NLM) is a large-scale research effort to develop knowledge-based tools and resources to compensate for differences in the way in which concepts are expressed in the field of biomedicine. Since 1990, a continually evolving set of UMLS knowledge sources has been released annually to the research community for experimentation and use in a wide range of applications (Lindberg, Humphreys, & McCray, 1993). Each year the knowledge sources are expanded and enhanced both with additional content and with additional research tools. More than 1000 institutions and individuals around the world use the UMLS in research and application.

The UMLS knowledge sources are the Metathesaurus, the Semantic Network, and the SPECIALIST lexicon. The Metathesaurus integrates vocabularies from the biomedical domain, and the Semantic Network is the network of general semantic categories, or types, to which all Metathesaurus concepts are assigned (McCray & Nelson, 1995). The heterogeneity in the nature, scope, and quality of the vocabularies that comprise the Metathesaurus makes it a particularly complex structure. A principal reason for developing the Semantic Network was to bring semantic coherence to this somewhat unwieldy structure. Together, the Metathesaurus and the Semantic Network express and classify a significant portion of the biomedical vocabulary.

The SPECIALIST lexicon and related lexical programs, which have been developed for natural language processing applications, are UMLS resources for managing the high degree of linguistic variation in natural language and in the terminologies themselves (McCray, Srinivasan, & Browne, 1994). The lexicon and lexical programs, which contain and manage not only biomedical terminology, but also a good portion of the general English vocabulary, may be used together with the other UMLS knowledge sources, but they may also be used independently in natural language processing applications. The UMLS Knowledge Source Server makes all UMLS resources available over the Internet through a Web-based interface, as well as through an application programming interface (McCray, Razi, Bangalore, Browne, & Stavri, 1996).

Building and revising the UMLS knowledge sources on an annual basis is a labor intensive process. A combination of automated and semi-automated methods is used, and this is followed by human review. Since human review is subject to human error, we have developed a variety of automated techniques to validate the data, and we correct errors as they are detected. Several investigators have explicitly analyzed the correctness, completeness, and usefulness of the UMLS (Bodenreider et al., 1998; Cimino, 1998; Pisanelli, Gangemi, & Steve, 1998; Srinivasan, 1999). Cimino, for example, has developed and implemented methods to "audit" the UMLS, using a number of interesting semantically-based techniques. The comments and analyses of those who actively use the UMLS have lead to significant improvements in the knowledge sources, and inform us as we continue to extend them.

2. CONCEPTS, CATEGORIES, AND RELATIONSHIPS

2.1 UMLS Metathesaurus

The most extensive of the three UMLS knowledge sources is the Metathesaurus. The first edition in 1990 contained approximately 30,000 concepts, representing a handful of vocabularies. The eleventh edition in 2000 contains over 730,000 concepts, representing more than 1,500,000 strings in over fifty vocabularies. These vocabularies include broad coverage biomedical terminolgies, such as the NLM's Medical Subject Headings (MeSH), disease specific terminologies such as the National Cancer Institute's PDQ vocabulary, drug terminologies such as the National Drug Data File, and medical specialty vocabularies such as the Classification of Nursing Diagnoses and the Current Dental Terminology. The vocabularies have in most cases been developed for differing purposes. MeSH, for example, is a thesaurus that is used to index the biomedical literature. The World Health Organization's International Classification of Diseases (ICD) was originally developed for the analysis and comparison of morbidity and mortality data throughout the world. In addition, however, a clinical modification of this terminology is used broadly in the U.S. for hospital and office visit billing purposes. Clinical Terms Version 3 (formerly the Read Codes) is a comprehensive clinical vocabulary that is currently used throughout the British health care system. Other terminologies have been developed for use primarily in computer-based patient records and in hospital information systems. In addition to differences in purpose and scope of coverage, the terminologies differ widely in size, with some having hundreds of terms, and others having tens of thousands. Some of the vocabularies are simply lists of terms, sometimes including synonyms, while others are fully structured thesauri (NISO, 1993), whose terms are interrelated in a variety of ways, and at least one source claims to be an ontology for a well-defined subdomain of medicine. Occasionally the relationships between the terms in the vocabulary are made explicit, but most often they are implicit.

Each Metathesaurus concept may be thought of as a cluster of synonyms. As a new vocabulary is added, lexical and other techniques are used to map it into Metathesaurus concepts. Thus, for example, if the concept "otitis media" already exists in the Metathesaurus and if a new vocabulary contains the term "middle ear infection", then as the new vocabulary is integrated, "middle ear infection" will become part of the synonym class that defines the concept "otitis media".

Metathesaurus concepts are related to multiple other concepts through a small number of broadly defined relationships. These include the 'child' and 'parent' relationships that are found in the constituent vocabularies. Additionally, during Metathesaurus construction, a pair of concepts may be identified as being related by a 'narrower than', 'broader than', or 'other' (saliently related to) relationship. Further, in some cases, the precise nature of the relationship is made explicit by adding a relationship "attribute" label. For example, if a pair of concepts is in the child/parent relationship, the attribute might be 'isa', or it might be 'part of'. This latter could be true, for example, in an anatomy hierarchy. Or, it might be that an 'other' relationship can be made more precise. For example, "middle ear" would be related to "otitis media" by the 'other' relationship in the Metathesaurus, but more specifically, it might also be given the attribute 'location of'. Currently, only about 25% of Metathesaurus concepts have specific relationship attributes. These attributes correspond to relationships specified in the UMLS Semantic Network. Finally, many concepts are inter-related because they co-occur in MEDLINE® bibliographic citation records. In the study we describe in Section 3 below, we take advantage of all of these inter-concept relationships as we build a more robust conceptual structure for the use of the UMLS through the Semantic Network.

As the Metathesaurus is built, the content and structure of each of the source vocabularies is preserved, and additional information is added at the concept level. This process, which is done annually, is a combination of computational techniques and human review. The SPECIALIST lexical tools are used to suggest likely related terms and these are then reviewed for correctness by UMLS editors, all of whom are specialists in the medical domain. The editors add a variety of information, including additional synonyms if these are available, and they add relationships to existing concepts. They categorize the concepts by assigning to each of them one or more semantic types from the Semantic Network. They review definitions and make sure that the cluster of synonyms accurately reflects the meaning of the concept.

Figure 1 is a sample of the type of information that is available for Metathesaurus concepts. Searching the UMLS Knowledge Source Server for the term "nearsightedness", would retrieve the information shown in the figure. The top left shows basic concept information, including the preferred concept name "myopia" and the concept unique identifier (C0027092). As noted, each concept is assigned a semantic type from the Semantic Network, and in this case it is "Disease or Syndrome". Further, since a definition is available, this is also displayed.

The synonyms that make up a concept are often drawn from multiple vocabularies. For this concept, this includes two versions of the International Classification of Diseases (ICD), the Medical Subject Headings (MSH), the Systematized Nomenclature of Medicine

 **UMLS Knowledge Source Server**

2000 release   http://umlsksnlm.nih.gov/

BASIC CONCEPT INFORMATION	RELATED CONCEPTS
Concept Name : Myopia **UI**: C0027092 **Semantic Type** : Disease or Syndrome **Definition** (MSH2000): A refractive error in which [...]. It is also called nearsightedness because the near point is less distant than it is in emmetropia with an equal amplitude of accommodation. **Synonyms** : • Nearsightedness • Error, refractive, myopia • SHORT SIGHTEDNESS • Near sighted • near vision **Sources** : ICD10, ICD2000, LCH90, MSH2000, PSY94, RCD99, SNM2, SNMI98, CCPSS99, COS92, CST95, DXP94, WHO97, AOD95, BI98, CSP98 **Other Languages** : • MYOPIE - French • Myopie - German • Kurzsichtigkeit - German • VISTA CURTA - Portuguese • CORTEDAD DE LA VISTA - Spanish	**Narrower Concepts** : • Degenerative progressive high myopia • Severe myopia **Broader Concepts** : • Refractive Errors • Ophthalmologic **Other related concepts** : • Vision Disorders • Abnormal vision • Eye problem • Blindness and vision defects **Co-occurring concepts** : • Cornea • Laser Surgery • Astigmatism • Keratotomy , Radial • Corneal Transplantation • Intraocular lens implant device • Postoperative Complications • Refraction, Ocular • Hyperopia • Eye • Visual Acuity [...]
ANCESTORS	DESCENDANTS

MSH2000
   Diseases (MeSH Category)
    Eye Diseases
     Refractive Errors
     Myopia
 	
RCD99
   Clinical findings
    Disorders
     Ophthalmological disorder
     Disorder of refraction and
             accommodation
     Disorder of refraction
     Myopia
	MSH2000
	Myopia
	<no children>
	
	RCD99
	Myopia
	• Malignant myopia
	• Pathological myopia
	• Simple myopia

Figure 1. Concept information in the UMLS for "myopia"

(SNM), and the World Health Organization (WHO) adverse drug reaction terminology. Translations in several languages are also available. The lower portion of the figure shows the hierarchical contexts of the concept in the various vocabularies in which it appears. In MeSH, myopia is a child of the term "refractive errors", while in the British Read (RCD) vocabulary it is a child of the term "disorder of refraction". (These two parent terms, while expressed somewhat differently, are actually the same concept and are represented as such in the Metathesaurus.) The top right of the figure shows that myopia is related to several other concepts in the Metathesaurus. Two concepts, "degenerative progressive high myopia" and "severe myopia", are listed as being narrower in meaning than myopia, and two are broader in meaning. Through the explicit 'other' relationship, myopia is closely related to several additional concepts, including "abnormal vision" and "eye problem". It frequently co-occurs with many other concepts in MEDLINE. Interesting implicit relationships hold between these co-occurring concepts. One might, for example, surmise that myopia 'has location' eye, 'is treated' by laser surgery, and 'affects' visual acuity. These co-occurring concepts are indicators of what is being written about in the biomedical literature, but, perhaps more interesting in this context, they are indicators of powerful associations among biomedical concepts, creating a latent semantic space for the domain (Landauer & Dumais, 1997).

2.2 UMLS Semantic Network

The UMLS Semantic Network has in common with most semantic networks that it consists of a collection of basic semantic types, which are the nodes in the network, and a set of relationships, which are its links (Brachman, 1979; Greenhill & Venkatesh, 1998; Lehmann, 1992; Quillian, 1968; Ruan, Burkle, & Dudeck, 2000; Sowa & Borgida, 1991; Woods, 1985). Quillian is most often credited with first developing the notion of semantic networks, and his work, though actually developed in the context of a psychological theory, heavily influenced subsequent work in knowledge representation. Brachman was among the first to critically evaluate the formal semantics of semantic networks and to suggest the ways in which networks might be used to build knowledge representation languages. More recent work, such as that of Greenhill & Venkatesh and Ruan et al., has emphasized the power of semantic networks to navigate complex knowledge spaces, particularly through robust visualization tools.

The UMLS Semantic Network was developed in order to provide a high level semantic structure for organizing the biomedical domain. It has the potential, through its 134 semantic types and 54 relationships, to simplify and bring coherence to a very large semantic space. There are two basic type hierarchies, one for entities and the other for events. Semantic types for organisms, anatomical structures, chemicals, concepts or ideas, behaviors, and physiologic and pathologic functions are included. There are also two categories of relationships. The first is the 'isa' relationship and the other comprises the non-hierarchical associative relationships. These latter are divided into five additional categories, including physical, spatial, functional, temporal, and conceptual relationships. (See table 1 for the full list of types and relationships.) While many of the semantic types are specific to the biomedical domain, most of the relationships are equally applicable in domains outside of medicine.

Entity
  Physical Object
    Organism
      Plant
        Alga
      Fungus
      Virus
      Rickettsia or Chlamydia
      Bacterium
      Archaeon
      Animal
        Invertebrate
        Vertebrate
          Amphibian
          Bird
          Fish
          Reptile
          Mammal
            Human
    Anatomical Structure
      Embryonic Structure
      Anatomical Abnormality
        Congenital Abnormality
        Acquired Abnormality
      Fully Formed Anatomical Structure
        Body Part, Organ, or Organ Component
        Tissue
        Cell
        Cell Component
        Gene or Genome
    Manufactured Object
      Medical Device
      Research Device
      Clinical Drug

[Entity] (continued)
[Physical Object] (continued)
    Substance
      Chemical
        Chemical Viewed Functionally
          Pharmacological Substance
            Antibiotic
          Biomedical or Dental Material
          Biologically Active Substance
            Neuroreactive Substance or Biogenic Amine
            Hormone
            Enzyme
            Vitamin
            Immunologic Factor
            Receptor
          Indicator, Reagent, or Diagnostic Aid
          Hazardous or Poisonous Substance
        Chemical Viewed Structurally
          Organic Chemical
            Nucleic Acid, Nucleoside, or Nucleotide
            Organophosphorous Compound
            Amino Acid, Peptide, or Protein
            Carbohydrate
            Lipid
              Steroid
              Eicosanoid
          Inorganic Chemical
          Element, Ion, or Isotope
      Body Substance
      Food

Table 1a. The UMLS Semantic Network: 'isa' relations between semantic types

[Entity] (continued)
  Conceptual Entity
    Idea or Concept
      Temporal Concept
      Qualitative Concept
      Quantitative Concept
      Functional Concept
      Body System
    Spatial Concept
      Body Space or Junction
      Body Location or Region
      Molecular Sequence
        Nucleotide Sequence
        Amino Acid Sequence
        Carbohydrate Sequence
      Geographic Area
  Finding
    Laboratory or Test Result
    Sign or Symptom
  Organism Attribute
    Clinical Attribute
  Intellectual Product
    Classification
    Regulation or Law
  Language
  Occupation or Discipline
    Biomedical Occupation or Discipline
  Organization
    Health Care Related Organization
    Professional Society
    Self-help or Relief Organization
  Group Attribute
  Group
    Professional or Occupational Group
    Population Group
    Family Group
    Age Group
    Patient or Disabled Group

Event
  Activity
    Behavior
      Social Behavior
      Individual Behavior
    Daily or Recreational Activity
    Occupational Activity
      Health Care Activity
        Laboratory Procedure
        Diagnostic Procedure
        Therapeutic or Preventive Procedure
      Research Activity
        Molecular Biology Research Technique
      Governmental or Regulatory Activity
      Educational Activity
    Machine Activity
  Phenomenon or Process
    Human-caused Phenomenon or Process
    Environmental Effect of Humans
    Natural Phenomenon or Process
    Biologic Function
      Physiologic Function
        Organism Function
          Mental Process
        Organ or Tissue Function
        Cell Function
        Molecular Function
          Genetic Function
      Pathologic Function
        Disease or Syndrome
          Mental or Behavioral Dysfunction
          Neoplastic Process
        Cell or Molecular Dysfunction
        Experimental Model of Disease
    Injury or Poisoning

Table 1a. The UMLS Semantic Network: 'isa' relations between semantic types—Cont.

isa	[associated with]
associated with	[functionally related to]
physically related to	performs
part of	carries out
consists of	exhibits
contains	practices
connected to	occurs in
interconnects	process of
branch of	uses
tributary of	manifestation of
ingredient of	indicates
spatially related to	result of
location of	temporally related to
adjacent to	co-occurs with
surrounds	precedes
traverses	conceptually related to
functionally related to	evaluation of
affects	degree of
manages	analyzes
treats	assesses effect of
disrupts	measurement of
complicates	measures
interacts with	diagnoses
prevents	property of
brings about	derivative of
produces	developmental form of
causes	method of
	conceptual part of
	issue in

Table 1b. The UMLS Semantic Network: 'isa' relations between relationships

Figure 2 shows a partial view of the Semantic Network, illustrating the kinds of relations that exist. As an example, note that anatomical structure is 'part of' an organism, an organism attribute is a 'property of' an organism, biologic function is a 'process of' an organism, and a plant 'isa' organism. By transitivity, a human also 'isa' organism.

The Semantic Network has a graph structure. Hierarchical (isa) relationships are organized in a single tree structure whereas associative relationships link semantic types from various levels of the hierarchical structure.

A range of information is provided for each semantic type. Each type is assigned a unique identifier and also a number that places it in the hierarchy of semantic types. For example, the unique identifier of "Experimental Model of Disease" is T050, and its tree number is B2.2.1.2.3, and it is a child of the semantic type "Pathologic Function" (B2.2.1.2). A definition is given for each type, and several examples of concepts to which this type may be assigned are also provided. In this case, the definition is "A representation in a non-human organism of a human disease for the purpose of research into its mechanism or treatment." A usage note assists those who are assigning semantic types to

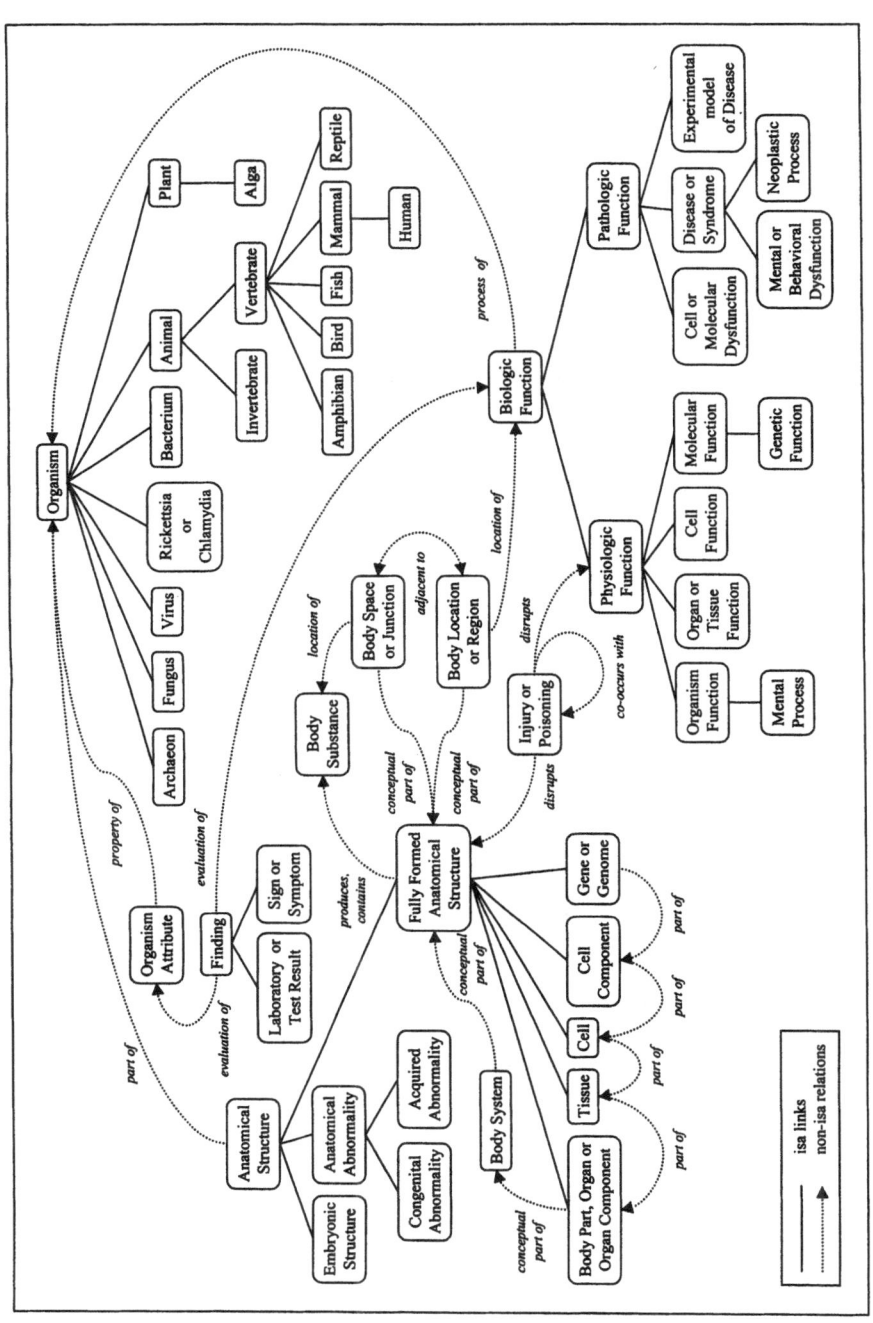

Figure 2. The UMLS Semantic Network (partial representation)

Metathesaurus concepts. In addition, through the UMLS Knowledge Source Server, a UMLS editor may at any time see the other concepts to which this semantic type has been assigned. "Experimental Model of Disease" has been assigned to fifty-three concepts in the current Metathesaurus, including "avian leukosis" and "experimental melanoma".

Analogously, information for relationships includes unique identifiers, the name of the relationship and its inverse, a tree number that places it in the hierarchy of relationships, a definition, and the pairs of semantic types it links. For example, the relationship 'treats' is defined as, "Applies a remedy with the object of effecting a cure or managing a condition." Its tree number is R3.1.2, and its parent in the relationship hierarchy is 'affects' (R3.1), which is in the 'functionally related to' (R3) hierarchy. The treats relationship links, for example, pathologic functions and injuries to pharmacologic substances, therapeutic procedures, and medical devices. The relationships are stated between high level semantic types and are generally inherited by all the descendants of those types. In this case, then, drugs are linked by the 'treats' relationship not only to pathologic functions, but also to all of the descendants of pathologic function, including diseases, mental or behavioral dysfunctions, and neoplastic processes. It is important to note that the relationships link semantic types to each other, but they do not directly link concepts to one another.

The UMLS Semantic Network has been explored by a number of researchers (Carenini & Moore, 1993; Gu et al., 2000; Joubert, Miton, Fieschi, & Robert, 1995; Volot et al., 1993; Yu, Friedman, Rhzetsky, & Kra, 1999). The focus of some of the work has been to "reuse" the knowledge encoded in the Semantic Network and express it in a variety of different knowledge representation frameworks. For example, Joubert et al. and Volot et al. have reformulated the Semantic Network in the closely related conceptual graph theory, and Gu et al. have represented both the Metathesaurus and the Semantic Network in an object oriented database framework. Among others, Carenini & Moore and Yu et al. have experimented with the Semantic Network in conceptualizing smaller, more focused domains within the broader biomedical domain.

3. BUILDING THE CONCEPTUAL FRAMEWORK:
AN EXTENDED EXAMPLE

In order to illustrate the ability and the limits of the UMLS Semantic Network to provide a conceptual framework for the biomedical domain, we designed the following study. Starting from a given concept, we gathered the concepts that constitute its semantic neighborhood by exploiting a set of inter-concept relationships represented in the UMLS Metathesaurus. For each pair of related concepts from this set, we calculated the possible relationships between the concepts using the semantic links defined in the UMLS Semantic Network between the semantic types that had been assigned to these concepts. Besides revealing the semantic structure in this set of concepts, other expected results included qualifying broadly defined relationships in the Metathesaurus, assessing already defined ones, and, more generally, by enforcing semantic rules, detecting inconsistencies in the Metathesaurus or in the Semantic Network itself.

The top part of figure 3 represents the semantic types and the relationships between them, as defined in the Semantic Network. The Metathesaurus, a set of concepts linked by inter-concept relationships, is represented in the bottom part of the figure. The two structures are related by means of the semantic types assigned to the concepts in order to categorize them.   Therefore, inter-concept relationships can be inferred, validated, or rejected by comparison to the relationships defined between the semantic types assigned to the concepts.

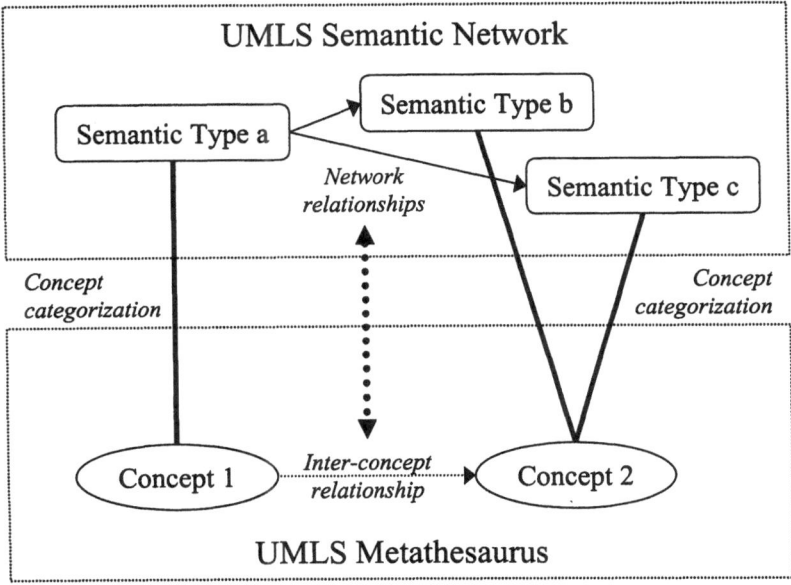

Figure 3. Defining the semantic structure for the domain.

3.1 Materials and Methods

3.1.1 Selecting a Set of Concepts for Experimentation

Starting from the concept "heart" (unique identifier: C0018787), we used the information available in the 1999 edition of the Metathesaurus to discover the concepts related to it:

- Concepts that are hierarchically related to "heart" through the 'parent', 'broader than', 'child', and 'narrower than' relationships were extracted. Hierarchically related concepts were extracted all the way to the top and bottom of the hierarchy, so that the set contains all the ancestors and descendants of "heart".
- Concepts that are related to "heart" through the 'other' (associative) relationship were extracted. In this case, only concepts directly associated with "heart" were selected.
- Concepts co-occurring with "heart" at least four times in MEDLINE were extracted. This selection represents 90% of all co-occurrences for "heart".

Using this methodology, we discovered 3764 closely related concepts. Figure 4 shows the partitioning of the concepts with respect to their relationship to "heart". Most concepts are related by only one type of relationship: Sixty-six are ancestors, 1952 are descendants, 242 are related through the 'other' relationship, and 1412 are co-occurring concepts. Some concepts are related by multiple relationships: One ancestor and twenty-seven descendants are also in the 'other' relationship to "heart". Two ancestors and thirty-two descendants also co-occur with "heart". Finally, twenty-five concepts in the 'other' relationship also co-occur with "heart", and five concepts simultaneously co-occur with, are descendants, and are in the 'other' relationship to "heart". These latter two groups are not shown in the figure.

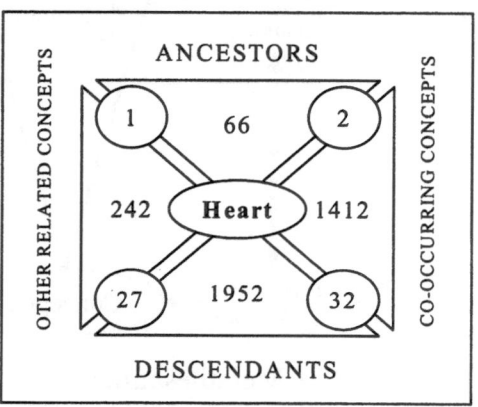

Figure 4. Origin of the concepts in the set of concepts related to "heart"

3.1.2 Preparing the Semantic Links

In order to facilitate knowledge processing, the UMLS provides a file of all semantic links resulting from the transitive closure of the Semantic Network graph. For example, no direct link is specified between the two semantic types "Disease or Syndrome" and "Body Part, Organ, or Organ Component". The 'location of' relation, however, may be inferred from the link between "Biologic Function" and "Fully Formed Anatomical Structure" since "Disease or Syndrome" is a descendant of "Biologic Function" and "Body Part, Organ, or Organ Component" is a child of "Fully Formed Anatomical Structure". Figure 5 shows an example of some additional semantic links that are calculated by the transitive closure of the graph.

For practical purposes, we complemented the list of explicit links between semantic types using two pieces of information implicit in the Semantic Network:

- We systematically added a reflexive 'isa' link (e.g. "Disease or Syndrome" isa "Disease or Syndrome") to reflect inter-concept relationships such as "myocardial infarction" 'isa' "cardiovascular diseases".

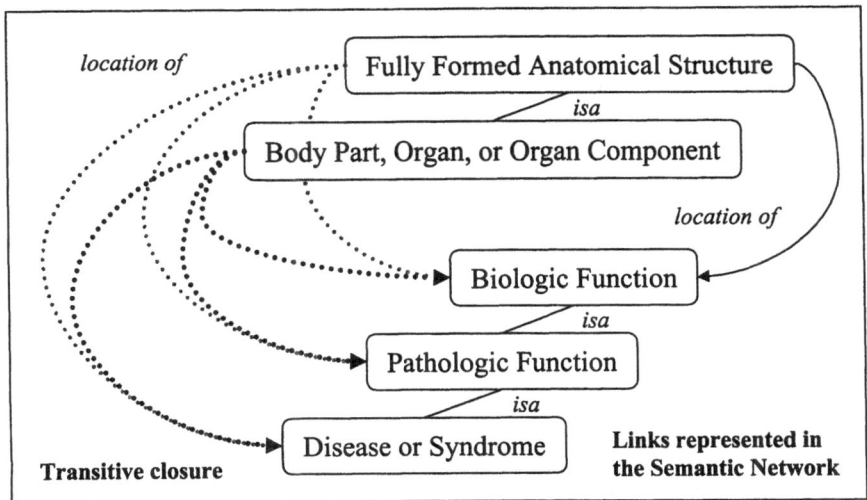

Figure 5. Additional semantic links calculated by transitive closure

- We added the symmetric inverse link for each pair of semantic types (e.g. "Body Part, Organ, or Organ Component" 'has location' "Disease or Syndrome", or "Disease or Syndrome" 'inverse isa' "Disease or Syndrome").

Once augmented, this resulted in 13,590 links (direct, inverse, and reflexive) among the 134 semantic types in the UMLS Semantic Network.

3.1.3 Revealing the Semantic Structure

Our set of 3764 concepts represents a total of 6894 pairs of related concepts. These numbers represent all the concepts directly related to "heart" as well as the concepts in the full set that are related to each other. The Metathesaurus may stipulate that a relationship exists between a pair of concepts, but it does not necessarily stipulate the precise nature ("attribute") of that relationship. To determine specific relationships between pairs of concepts, we followed several steps:

We abstracted away from the concepts themselves and compared their semantic types to the allowable Semantic Network links.

- If there was only one possible relationship for the semantic type pair, this one was tentatively chosen as the inter-concept relationship.
- If there were multiple possible relationships between the semantic type pair, then the Metathesaurus was checked for the original broad relationship between the concepts. In this case,
  - If the original broad relationship was a hierarchical one ('child', 'parent', 'broader than' or 'narrower than'), and if 'isa' was an allowable relationship between the semantic type pair, then 'isa' was tentatively chosen as the inter-concept relationship.

- If the original broad relationship was associative ('other') or if the concepts were derived from co-occurrence, then if there was only one allowable associative relationship between the semantic type pair, it was tentatively chosen.
- If there was more than one associative relationship, then it was not possible to make a tentative assignment, and the pair was ambiguous.
- If no relationship existed between the semantic type pair, then no inter-concept relationship could be inferred.

All tentative and ambiguous inter-concept assignments were then checked in the Metathesaurus to see if there were relationship attributes listed for the specific concept pair. In this case,

- If the attribute was compatible with an allowable relationship between the semantic type pair, it was chosen as the inter-concept relationship.
- If there was no attribute, or if the attribute was incompatible, the inter-concept relationship could not be resolved.

3.2 Results

Among the 6894 pairs of related concepts, we obtained the following results:

- In 4496 cases (65%), a semantic relation could be inferred unambiguously from the Semantic Network. The semantic relation inferred allowed us to determine inter-concept relationships whose attribute was not defined in 2515 of these cases, and to confirm the validity of the relationship attribute in 1981 of these cases.
- In 1491 cases (22%), multiple semantic links existed between the semantic types of the two concepts, leading to several possible attributes for these inter-concept relationships.
- In the remaining 907 pairs (13%), the inter-concept relationships represented a violation of the Semantic Network. In 372 pairs, there was no semantic link between the semantic types of the two concepts. In 415 pairs, the inter-concept relationship was not compatible with that of the corresponding Semantic Network relationship. Finally, in 120 pairs, the attribute of the inter-concept relationship was not compatible with the semantic relationships allowed between the semantic types of the two concepts.

Although the relevance of the semantic relations inferred from the Semantic Network is difficult to evaluate without systematic human review, their compatibility with the relationship attribute, when specified, might provide an estimate. In 89% of the cases, the semantic relationship inferred from the Semantic Network (unambiguously or not) is compatible with the relationship attribute.

Figure 6 shows a portion of the conceptual framework we built for cardiology. The most common relationships in this set of concepts are 'isa', organizing diseases, and 'location of', linking diseases or procedures to the corresponding "Body Part, Organ, or Organ Component". Various other associative relationships help structure the domain. The attribute of inter-concept relationships is accurately and unambiguously inferred in most cases. However, in some cases attributes are incorrectly inferred. For example, 'isa' is inferred between "heart" and "heart valves", while it should be 'part of'. Finally, the relationship of the sign "systolic murmur" to the disease "mitral valve insufficiency" cannot be selected automatically since three possible relationships are defined between their semantic types.

3.3 Discussion

The Semantic Network is able to suggest a semantic structure from a set of concepts in the Metathesaurus. Additional knowledge about inter-concept relationships is not technically required for inferring semantic relations from the Network. However, since the Semantic Network defines relationships between very broad categories, these may not hold between concepts to which these categories are assigned. For example, "heart" 'location of' "intracranial pressure" is valid according to the Semantic Network, since a "Body Part, Organ or Organ Component" is the location of an "Organ or Tissue Function". However, this is an incorrect inference. Therefore, the Semantic Network is best used in collaboration with the Metathesaurus, where the existence of a particular relationship, hierarchical or associative, between concepts is described, even though broadly. Using the methods we have described here, the Semantic Network is able to suggest the appropriate, more specific, relationship. The results of this study do not suggest an entirely automated approach for organizing biomedical concepts, in particular since about 20% of the semantic relationships could not be inferred unambiguously. However, the method can be used to suggest possible relationships to the human editors of the Metathesaurus. For example, the relationship of "heart" ("Body Part, Organ, or Organ Component") to "cobra venoms" ("Biologically Active Substance") may be either 'produces' or 'disrupted by'. A human editor will easily select the latter when prompted with this selection.

This method also helps detect discrepancies between the semantics of the Metathesaurus and the semantics expressed by the Semantic Network. Automatic detection helps limit the need for human review by focusing on conflicting relationships that violate the semantic rules. One major cause for such a violation is somewhat artificial: Concepts with an abstract semantic type (e.g. "Classification") may have related concepts having a concrete semantic type (e.g. "Body Part, Organ, or Organ Component"). These types would be unrelated in the Semantic Network. The relationship of "heart: general terms" to "right side of heart", for example, violates the Semantic Network for this reason. Another source of problems is that partonymic relationships (part of) are considered associative in the Semantic Network, while in many medical vocabularies they are used hierarchically. Frequently occurring semantic discrepancies may also help identify missing semantic links in the Semantic Network. For example, the relationship of "chest pain" to "thorax" violates the Semantic Network, since the 'location of' relationship has not been defined between a "Body Location or Region" and a "Sign and Symptom".

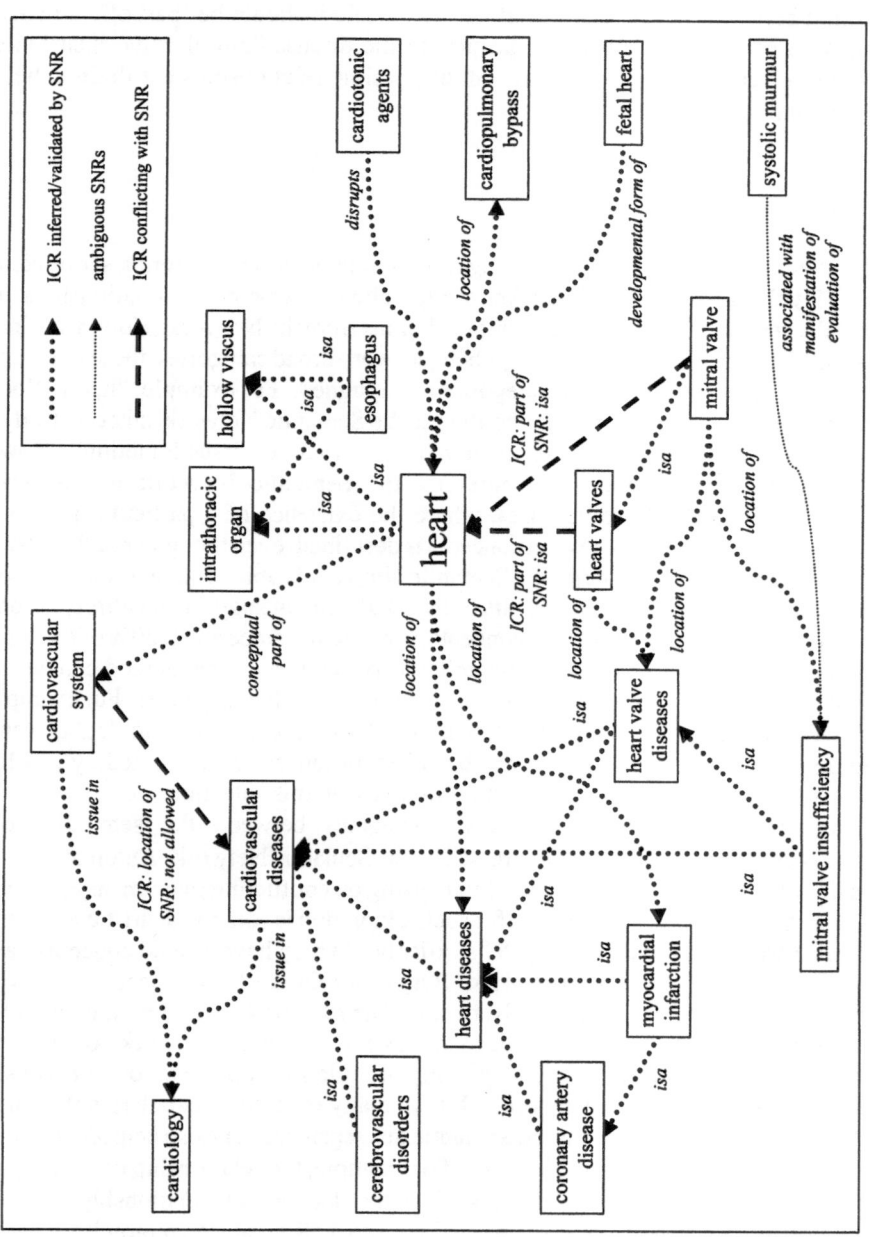

Figure 6. A portion of the conceptual framework for cardiology (attributes of the inter-concept relationships [ICRs] as suggested from semantic network relationships [SNRs])

4. CONCLUSIONS

In the UMLS each concept can be understood and defined by its relationships to other concepts and by the semantic category to which it belongs. This principle, called semantic locality, is an important organizing principle in the UMLS. The semantic structure of the UMLS consists of two related parts: the Semantic Network and the Metathesaurus. The Semantic Network provides a small number of strong semantic rules, by defining relationships among a small number of high-level semantic types. The Metathesaurus provides a large number of inter-related concepts. The semantic typing of all concepts in the Metathesaurus allows it to inherit the semantic rules provided by the Semantic Network. Together, the two knowledge sources define a large portion of the biomedical domain.

Ideally, inter-concept relationships in the Metathesaurus could be limited to representing factual knowledge, relying on the Semantic Network to interpret this knowledge. For example, it might be sufficient to know that "heart" is related to "heart diseases", if the nature of the relationship can be inferred unambiguously at a higher level. In fact, however, because the Semantic Network has a limited number of semantic types of coarse granularity for a domain as broad as biomedicine, it does not allow us to completely achieve such a goal. However, we have shown that the Semantic Network can be used to infer quite accurately the nature of the relationships between concepts in the Metathesaurus in a particular subdomain of medicine. The results suggest that this method can be applied more broadly to the Metathesaurus as a whole.

References

Bodenreider, O., Burgun, A., Botti, G., Fieschi, M., Le Beux, P., & Kohler, F. (1998). Evaluation of the Unified Medical Language System as a medical knowledge source. *Journal of the American Medical Informatics Association, 5*, 76-87.

Brachman, R. J. (1979). On the epistemological status of semantic networks. In N. Findler (Ed.), *Associative Networks: Representation and Use of Knowledge by Computers*, 3-50. New York: Academic Press.

Carenini, G., & Moore, J. D. (1993). Using the UMLS Semantic Network as a basis for constructing a terminological knowledge base: A preliminary report. *Proceedings of the 17th Annual Symposium on Computer Applications in Medical Care*, 725-729.

Cimino, J. J. (1998). Auditing the Unified Medical Language System with semantic methods. *Journal of the American Medical Informatics Association, 5*, 41-51.

Greenhill, S., & Venkatesh, S. (1998). Noetica: A tool for semantic data modelling. *Information Processing & Management, 34*, 739-760.

Gu, H. Y., Perl, Y., Geller, J., Halper, M., Liu, L. M., & Cimino, J. J. (2000). Representing the UMLS as an object-oriented database: Modeling issues and advantages. *Journal of the American Medical Informatics Association, 7*, 66-80.

Joubert, M., Miton, F., Fieschi, M., & Robert, J. J. (1995). A conceptual graphs modeling of UMLS components. *Medinfo, 8*, 90-94.

Landauer, T. K., & Dumais, S. T. (1997). A solution to Plato's problem: The latent semantic analysis theory of acquisition, induction, and representation of knowledge. *Psychological Review,* 104, 211-240.

Lehmann, F. (1992). Semantic networks. In F. Lehmann (Ed.), *Semantic Networks in Artificial Intelligence,* 1-50. Tarrytown, NY: Academic Press.

Lindberg, D. A., Humphreys, B. L., & McCray, A. T. (1993). The Unified Medical Language System. *Methods of Information in Medicine,* 32, 281-291.

McCray, A. T., & Nelson, S. J. (1995). The representation of meaning in the UMLS. *Methods of Information in Medicine,* 34, 193-201.

McCray, A. T., Razi, A. M., Bangalore, A. K., Browne, A. C., & Stavri, P. Z. (1996). The UMLS Knowledge Source Server: A versatile Internet-based research tool. *Proceedings of the 1996 AMIA Annual Fall Symposium,* 164-168.

McCray, A. T., Srinivasan, S., & Browne, A. C. (1994). Lexical methods for managing variation in biomedical terminologies. *Proceedings of the 18th Annual Symposium on Computer Applications in Medical Care,* 235-239.

National Information Standards Organization (NISO). (1993). *Guidelines for the Construction, Format, and Management of Monolingual Thesauri (Developed by the National Information Standards Organization. Approved August 30, 1993, by the American National Standards Institute).*

Pisanelli, D. M., Gangemi, A., & Steve, G. (1998). An ontological analysis of the UMLS Methathesaurus. *Proceedings of the 1998 AMIA Symposium,* 810-814.

Quillian, M. (1968). Semantic memory. In M. Minsky (Ed.), *Semantic Information Processing,* 227-270. Cambridge, MA: MIT Press.

Ruan, W., Burkle, T., & Dudeck, J. (2000). An object-oriented design for automated navigation of semantic networks inside a medical data dictionary. *Artificial Intelligence in Medicine,* 18, 83-103.

Sowa, J. F., & Borgida, A. (1991). *Principles of Semantic Networks : Explorations in the Representation of Knowledge.* San Mateo, CA: Morgan Kaufmann.

Srinivasan, P. (1999). Exploring the UMLS: A rough sets based theoretical framework. *Proceedings of the 1999 AMIA Symposium,* 156-160.

Volot, F., Zweigenbaum, P., Bachimont, B., Ben Said, M., Bouaud, J., Fieschi, M., & Boisvieux, J. F. (1993). Structuration and acquisition of medical knowledge. Using UMLS in the conceptual graph formalism. *Proceedings of the 17th Annual Symposium on Computer Applications in Medical Care,* 710-714.

Woods, W. (1985). What's in a link: Foundations for semantic networks. In R. J. Brachman & H. J. Levesque (Eds.), *Readings in Knowledge Representation,* 218-241. Los Altos, CA: Morgan Kaufmann.

Yu, H., Friedman, C., Rhzetsky, A., & Kra, P. (1999). Representing genomic knowledge in the UMLS semantic network. *Proceedings of the 1999 AMIA Symposium,* 181-185.

Chapter 12

Visual Analysis and Exploration of Relationships

Beth Hetzler
Pacific Northwest National Laboratory, Richland, WA, USA

Abstract:
Relationships can provide a rich and powerful set of information and can be used to accomplish application goals, such as information retrieval and natural language processing. A growing trend in the information science community is the use of information visualization—taking advantage of people's natural visual capabilities to perceive and understand complex information. This chapter explores how visualization and visual exploration can help users gain insight from known relationships and discover evidence of new relationships not previously anticipated.

1. INTRODUCTION

The chapters in this book describe many examples of rich and potentially very useful relationships in the information we access. As our analysis of these relationships becomes more sophisticated and our potential applications for such relationships multiply, one challenge is how to represent these relationships so that people can easily understand and use them. Our solutions to this challenge must depend on the particular task and needs of the person who will use them. A person who wants to start with a particular word and find a list of hyponyms and antonyms may be well served by a simple character-based interface. However, someone trying to understand a large number of interrelationships among many words probably needs an approach much more tuned to getting the "big picture."

Information visualization uses people's ability to quickly understand and derive meaning from visual scenes. Although it can be used to communicate relatively simple concepts, its real power is in communicating complex information. If someone were to describe all the relationships (e.g., spatial, similarity, comparative size, etc.) seen among the surrounding objects in a room, the description might be very lengthy, although the relationships can be visually perceived and understood very quickly—even allowing for a moment's pause to examine something in more detail or from a different angle. Thesauri are a compilation of many potentially complex relationships, usually presented textually. Bertrand-Gastaldy (1986) lists several potential benefits of using graphics in thesauri, including the ability to show the overall environment of a concept, the ability to show multiple diverse relationships, and easier use by people unfamiliar with indexing terms and conventions.

Information visualization relies on the capabilities of human vision, perception, and cognition, a combination that Hoffman (1998) refers to as "visual intelligence." For a number of reasons, this combination is ideally suited for conveying information:

- Vision is a highly sophisticated parallel-processing capability. We see many things at the same time, without needing to look at each one individually. Images can convey a lot of information in relatively little time.
- Visual objects can also attract our attention, making us focus directly on them. This can be particularly useful in identifying the unexpected or things that look "out of place."
- We are able to view a given scene at multiple levels of detail. We can look at the overall or abstract view, or we can concentrate on particular details.
- We perceive many attributes of objects visually, such as color, texture, size, and brightness. When used carefully, these can convey information, such as object characteristics.
- We perceive patterns in what we see, such as parallel or colinear lines, repeating shapes, etc. These can be used to convey relationships as well.
- Graphical images can help us remember better what we learn—for example, by providing an additional memory encoding (Ashcraft, 1989) and by reducing the amount of information to be stored.

These and other qualities of vision can be employed to help convey information that is known and also to help people discover new information. In both cases, visual exploration capabilities can provide additional value.

Many of us have seen examples where graphics can convey clearly and immediately what is difficult to pick out from a table of numbers. Many of us have also seen examples where graphics at best fail to clarify and at worst obscure the information. Tufte (1997) provides a compelling set of examples showing various possible ways to present the data about O-ring performance in space shuttle launches prior to the Challenger accident, with a wide range of effectiveness.

To design an effective visualization, it's important to first understand the characteristics of the user's tasks and of the information. Does the user want to get an overview, to identify outliers, to see relationships pertinent to a single object of interest, to identify groupings, to compare two sets of relationships? Are the relationships one-way or two-way? Do they have varying strengths? Are they discrete (such as same author) or continuous (such as temporal relationships)? All of these will influence the choice of how to visualize.

Equally important is a good understanding of how the human brain interprets visual input and how easily different approaches can be understood. Human vision research is a highly active field. Hoffman (1998) provides an interesting summary of what is currently known on how the brain constructs color, shape, motion, texture, and other characteristics of the scenes we see. Several empirical studies have been conducted on the relative effectiveness of graphical attributes, such as position, color, size, and shape for conveying object characteristics (for examples, see Christ, 1975 and Cleveland & McGill, 1984). The execution of such controlled studies depends on the selection of a specific task and particular user group, so their conclusions are not necessarily generalizable to all users and visualizations (Nowell, 1997). The bottom line is that while guidance about approaches and particular choices is available, the creation of good visualizations remains more an art than a science.

One technique is to base the visualization on concepts familiar to the user, by employing metaphors. Examples include a landscape metaphor (Wise et al., 1995), a butterfly metaphor (MacKinlay, Rao, & Card, 1995), and a river metaphor (Havre, Hetzler, & Nowell, 1999). Such visualizations seek to take advantage of people's previous experience to make it easier to learn and understand the information being portrayed. Lakoff and Johnson (1980) draw evidence from linguistic expressions and argue that certain metaphors are wired into our understanding of particular concepts—for example UP IS GOOD. Theoretically, visualizations that uphold such basic metaphors would be more intuitive than those that violate them.

Visual interaction—the ability for the user to directly change the visual presentation in some meaningful way—is an important part of most visualization systems. Common examples include abilities to:

- Pan, zoom, or rotate the display;
- Show more or less detail about some portion of the information;
- Dynamically change the mappings between information characteristics and graphical portrayal, e.g., document dates mapped to color or size;
- Filter the number or selection of objects shown; and
- Highlight selected objects across multiple windows.

Some systems place the majority of their focus on such exploratory capabilities (e.g., Ahlberg & Shneiderman, 1994).

This chapter presents a number of examples of how visualization techniques have been used for representation and exploration of relationships. Often, particular types of relationships can be presented in multiple ways; likewise, particular visualization approaches may be used to convey multiple types of relationships. For this chapter, a varied selection has been chosen to demonstrate how relations discussed in this book might be presented and explored. The organization is by relationship type:

- hierarchical,
- temporal,
- cause/effect,
- similarity,
- explicit links or citations, and
- lack of relationships.

A full survey of relationship visualization is beyond the scope of this chapter; rather, the goal is to provide examples to suggest a range of possibilities. For a broader intro-duction to the field, see Card's, MacKinlay's, and Shneiderman's *Readings in Information Visualization* (1999), Chen's *Information Visualization and Virtual Environments* (1999), or the annual *Proceedings of IEEE Information Visualization* (for example, Wills & Dill, 1998; Wills & Keim, 1999).

2. VISUALIZING HIERARCHIES

Hierarchical relationships are common in the world around us. Different flavors—hyponomic and meronomic—are discussed in Cruse (this volume) and Pribbenow (this volume). Organization charts and family trees are two familiar types of hierarchical visualizations. Bertrand-Gastaldy's survey of graphic displays in thesauri (1986) found that the most prevalent type of graphic was a "tree-like structure." Hierarchical relationships are transitive; thus only the immediate relationships are shown explicitly.

Hierarchies are commonly visualized with some form of tree-like graph. According to Lancaster (1972), one of the first thesauri to use graphics was the TDCK Circular Thesaurus (TDCK, 1963). It employed a circular tree structure, with the most general term at the center and successive levels of more specific terms shown grouped along concentric circles moving out from the center.

With a large and complex hierarchy, visibility and amount of detail can be a significant issue. A number of interactive hierarchy visualizations have been developed recently to help address this issue. They provide capabilities such as expanding and contracting the tree levels, rearranging the spatial layout, and filtering to focus attention on a particular tree lineage. Figure 1 shows Hearst's and Karadi's (1997) interactive three-dimensional workspace called Cat-a-Cone, designed to support interactive searching and browsing in the context of a hierarchical index. In this figure, the user is searching through a set of medical abstracts, indexed with MeSH (Medical Subject Headings) categories. The three-dimensional tree structure is called the ConeTree. If a user specifies a particular term, the corresponding cone at each level of the tree can rotate such that the parentage of that term is in the front and is highlighted. The dynamic movement helps the user keep track of the context of the whole structure while focusing on the detail at hand. In the foreground, the details of the retrieved selection are shown using a workbook-style presentation.

The Disc Tree (Jeong & Pang, 1998) visualization uses the same principles as the ConeTree, but puts each hierarchy level in a flat disk to reduce potential occlusion problems. Some approaches, such as the hyperbolic tree (Lamping & Rao, 1996) and 3D hyperbolic tree (Munzner, 1997), use an alternate mathematical representation to keep the tree within a fixed space while enlarging the portion of interest, again relying heavily on user interaction.

A very different method for depicting hierarchies is given by space-filling methods. These methods use spatial layout and proportion, not only to show the hierarchical structure, but also to convey some attribute of the hierarchy members. Figure 2 provides an example of a space-filling method called a TreeMap, where the hierarchy is the organization of a particular library collection based on the Dewey Decimal system (Johnson & Shneiderman, 1991). The top levels of the hierarchy are shown as vertical rectangles, dividing the space into 10 blocks, where the size of each represents the number of books available in that category. To show the next level, each vertical block is divided into horizontal rectangles, with size again representing the number of books, and so on. Other attributes can be shown using color or texture. Here the redder areas indicate more frequent use, while the blue areas are less frequently used. One compromise between the TreeMap space-filling approach and the treelike approaches shows the hierarchy levels as successive circular bands, as in the Sunburst visualization (Stasko & Zhang, 2000) with the hierarchy root at the center and each successive level radiating out from it.

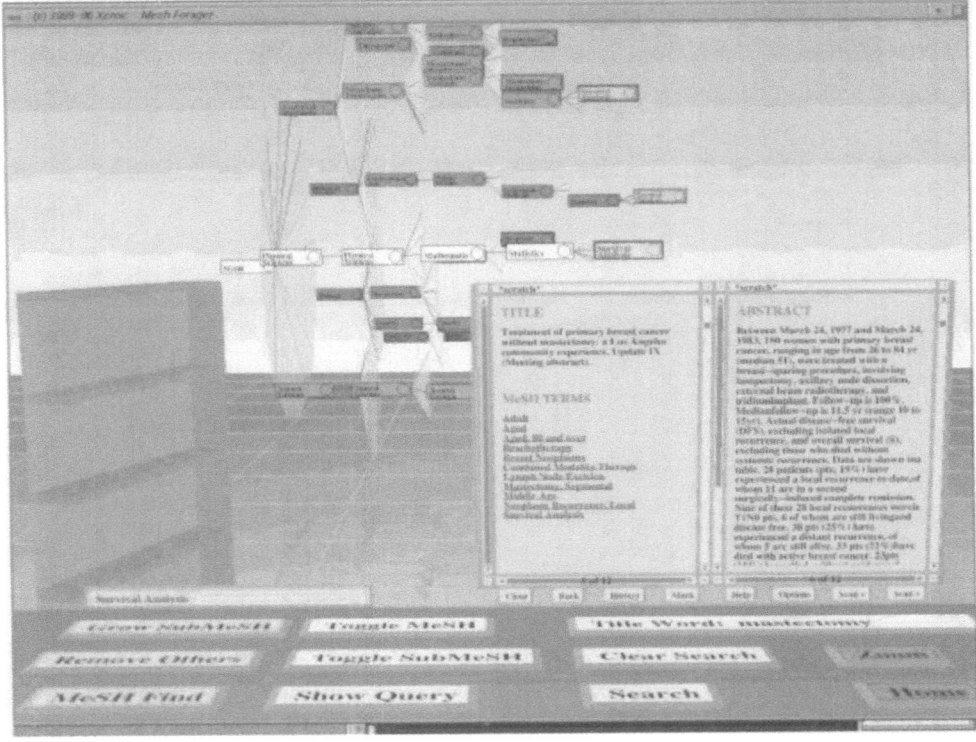

Figure 1. The Cat-a-Cone combines an animated hierarchy with a detail view and other task supports. Printed with permission of Marti Hearst.

3. TEMPORAL RELATIONSHIPS

On one hand, temporal relationships can be viewed as an instance of ordinal relationships. Documents' publication dates provide an attribute that can be presented in many forms (e.g., position, color) to show the sequence. However, in many contexts, the aspect of *duration* must also be considered. For example, in patent analysis, the time between patent application and patent issuance may be important. In studying the influences among authors, the duration and overlap of their productive periods may be important.

Figure 3 shows a system, called TmViewer, displaying the reigns of English rulers along with related events (Kumar & Furuta, 1999). Such a view might be used to analyze relationships among historical events. In this figure, time goes from left to right along the x-axis. The y-axis distinguishes the events into various categories. Parent-child relationships among the rulers can be shown as lines. A similar visual approach is taken by the LifeLines system developed jointly by the University of Maryland and IBM (Plaisant et al., 1998 and Plaisant, Milash, Rose, Widoff, & Shneiderman, 1996) and used to visualize medical records and juvenile criminal records.

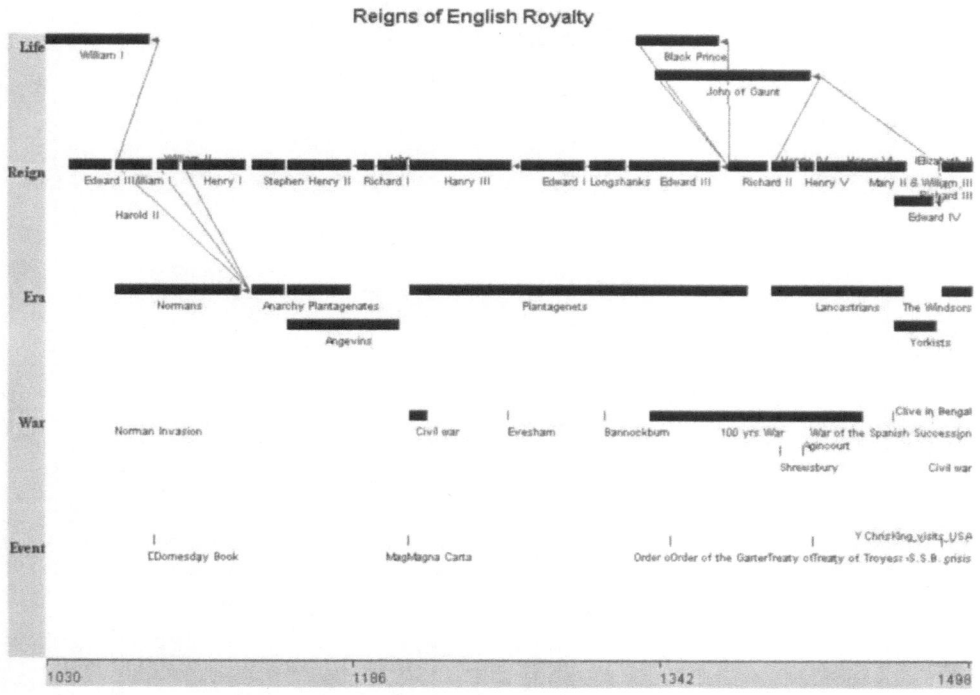

Figure 3. TmViewer showing temporal relationships among English historical events. Kumar & Furuta © 1999 Association for Computing Machinery, Inc. Reprinted by permission.

Another user task might be to understand how themes gain and lose strength over time, perhaps in the life of an author, perhaps as their popularity changes in literature. ThemeRiver is a prototype system that visualizes thematic variations over time across a collection of documents (Havre, Hetzler, & Nowell, 2000). As shown in figure 4, the "river" flows through time from left to right. Themes or topics are represented as colored "currents" flowing within the river that narrow or widen to indicate decreases or increases in the strength of a topic in associated documents at a specific point in time. The river is shown within the context of a timeline and a corresponding textual presentation of external events.

The DIVA system (Mackay & Beaudouin-Lafon, 1998) uses the approach shown in figure 5 to show how particular measured values change in relation to the temporal flow and left edges of the planar face containing the video. In this example, we see that the D measure is "on" at the time corresponding to the current video frame, as shown by the dark square just above the D on the top edge of the face. The "past" values are progressively further away from these edges; the D square seems to bend up and over the edge of the plane and continue, showing that this measure was also on for the previous video frame. The slot along the bottom edge and directly below the D is also dark, showing that the D measure will be on for the next video frame as well. As the video progresses, the measured

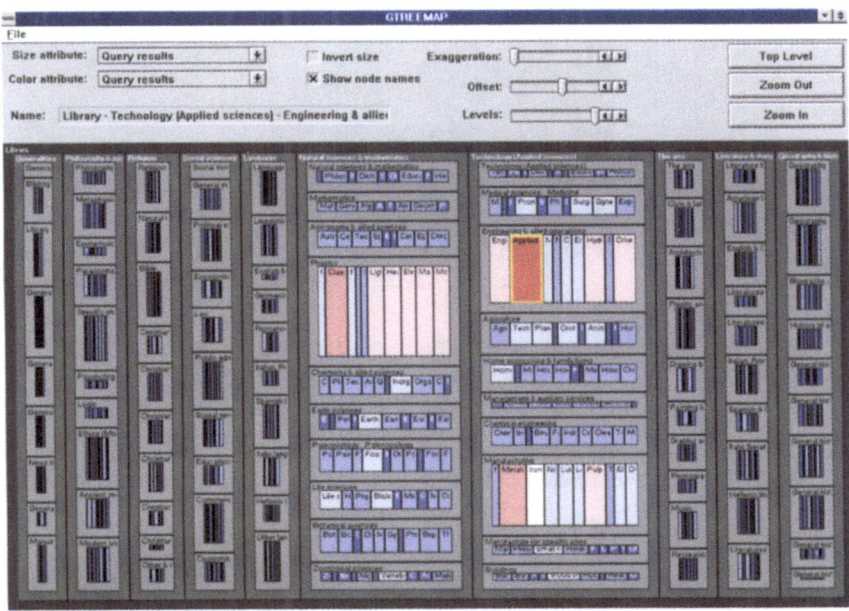

Figure 2. Tree map of a library collection based on the Dewey Decimal system.
Courtesy of University of Maryland.

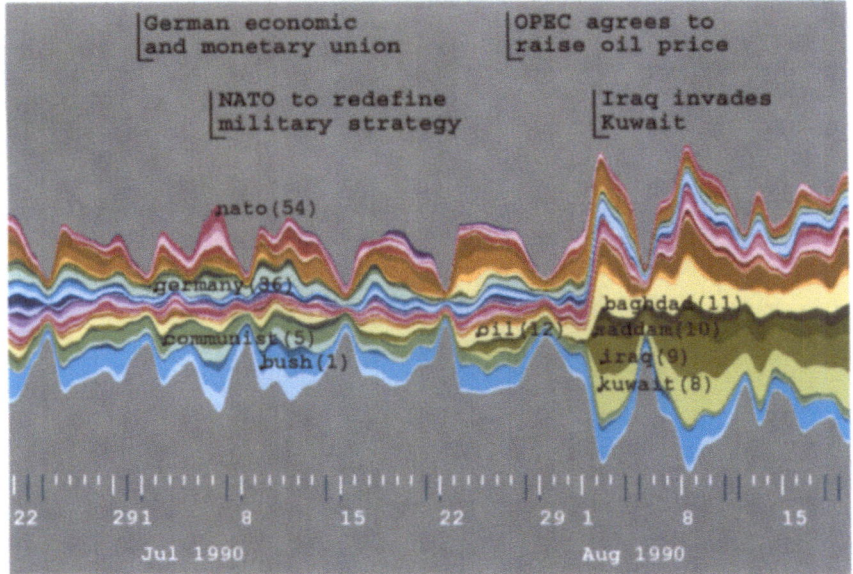

Figure 4. ThemeRiver displays changing strength values over time and in the context of events.

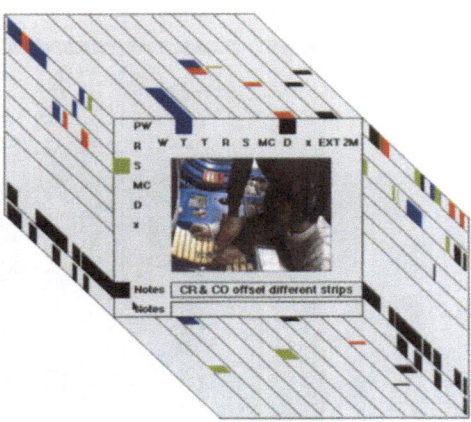

Figure 5. The DIVA system relates measured values to video action.
Mackay & Beaudouin-Lafon © 1998 Association for Computing Machinery.
Reprinted by permission.

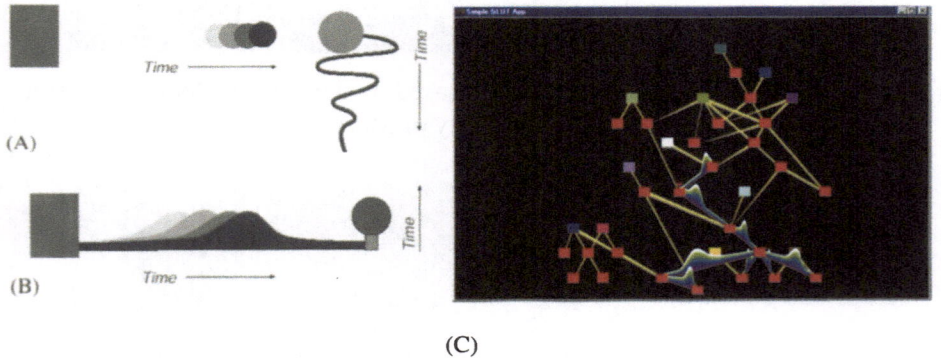

(C)

Figure 6. Examples of animation methods used to denote causality. Images provided by Colin
Ware, University of New Brunswick and Eric Neufeld, University of Saskatchewan.

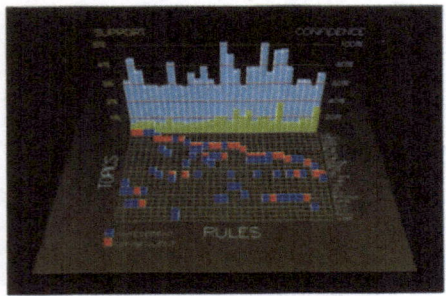

Figure 7. Visualization of association rules relating antecedents to consequents with
multiple causes.

values seem to approach from the lower and right directions, jump to the planar face containing the video, and then recede toward the top and left.

These visualizations provide an example of relationship discovery. They explicitly present only the temporal relations. However, in exploring these relations, users may be able to discover or infer additional relations. For example, ThemeRiver may help users discover unanticipated relationships between themes and external events. The DIVA system may help users notice that two particular measured values show nearly identical patterns during portions of the video.

4. VISUALIZING CAUSALITY

Khoo, Chan, & Niu (this volume) provides an overview of the concept of cause and effect from a variety of perspectives. One common method for visualizing such causal relationships is directed graphs. Each entity is represented by a node, and the relationship is shown as an arrow from the cause to the result. Individual instances are easily represented in this way; however, large sets of causal relationships can yield complex graphs, especially when one-to-many or many-to-many relationships are involved. Additional information, such as size of the effect, confidence, etc., may be shown with graphical attributes. For example, the confidence of a relationship may map to the thickness, opacity, or color of the edge representing it.

Animation has also been proposed as a method for visualizing causality. Experiments show that people infer causality when they see objects moving at specific times in relation to each other (Michotte, 1963). A real world analogy is the motion of billiard balls: Immediately after a first ball hits a second, the second ball begins to move; we infer a cause-effect relationship. Figure 6 shows examples of animation methods under investigation to denote causality (Ware, Neufeld, & Bartram, 1999). Figure 6a is based on the billiard ball example; the two related entities are represented as the rectangle on the left and the ball on the right. A smaller ball moves from the rectangle to the larger ball, which then oscillates back and forth, dampening over time. In Figure 6b, the relationship is shown with a wave that flows from the rectangle to the ball, which bounces up in response. These causal instances can be strung together in a graph layout that is then animated to show the paths of causality. Figure 6c is a single frame from such an animation, representing multiple causal relations with waves moving between nodes.

A related concept in the field of data mining is called Association Rules. Here the goal is to find and display instances where the presence of a set of entities (called the antecedent) implies that an additional entity (called the consequent) will also be present. An example might be "grocery shoppers who buy diapers and baby powder also tend to buy baby oil." While not strictly cause/effect relations, such rules have similar properties for visualization purposes. Directed graphs may be used to show the rules discovered. Another approach that has been used for such relationships is a matrix, where the antecedents are shown along one side; the consequents are shown along the other side; and the intersections relate the two, as in SGI's MineSet (1999).

A recent variation on this approach is given in figure 7 (Wong, Whitney, & Thomas, 1999). Each rule is shown as a vertical column in the matrix. The antecedents in the rule are shown in one color and the consequents are shown in another color. The bar graphs along the back denote how strongly the antecedent and consequent are present in the collection and how likely the rule is to hold.

5. VISUALIZING SIMILARITY

Similarity comes in various flavors, including synonymy, spatial nearness, metonymy, and entailment. Another kind of similarity is that found in object attributes. For example, words can have similar parts of speech, similar conjugations (regular or irregular), or similar language roots. Documents may have common authors, common sources, or common sizes.

A key part of many information visualization approaches is their methods for portraying similarity. Sometimes the goal is to convey different aspects of similarity to the user. Sometimes the goal is to discover similarity. A common method for visualizing similarity relationships over a large set of objects is to use spatial proximity. Each object is shown as a small graphical icon, such as a dot or ball. These icons are arranged in a two-dimensional or three-dimensional space where icons near each other represent similar objects, and icons that are far apart represent dissimilar objects. Spatial proximity has attractive qualities; it maps closely to our intuition and conveys both local relationships and global ones. Various calculation methods have been used to create proximity-based visualizations, including multidimensional scaling (Seber, 1984), Kohonen self-organizing maps (Kohonen, 1997), spring-based modeling (Sprenger, Gross, Bielser, & Strasser, 1998), and principal components (Seber, 1984).

An example is shown in figure 8. Here a search result containing 1287 documents is visualized using a Kohonen map layout (Lin, 1997). The documents are shown as dots and arranged in a two-dimensional layout. Area boundaries are drawn and labeled to help the user interpret the natural groupings that emerge. Terms that tend to occur together in the documents are shown as neighboring areas in the map.

Visualizations built on proximity displays often provide additional information through the use of metaphors or graphical features. The ThemeView visualization uses a landscape metaphor (see fig. 9), where proximity indicates semantic similarity or relatedness among documents and the height of the hills indicates thematic strength (Wise et al., 1995).

Risch et al. (1997) describes the STARLIGHT system, which uses a three-dimensional spatial layout to show similarity and adds the ability to show linkages, such as source locations for the documents or common locations mentioned within them (see fig. 10). Such a presentation might allow a user to discover that most research on a particular topic is concentrated in two locations. The Bead system described by Brodbeck, Chalmers, Lunzer, & Cotture (1997) uses icon color and shape to portray other object attributes.

Another common method for portraying similarity is to use graphical attributes. Users perceive that the icons have a common shape or color and conceptually can make the mapping between the graphical similarities and the object similarities being portrayed, perhaps with the help of a legend. An example is given in figure 11, where the strong themes in a document are shown as knobby protrusions from a spherical shape (Rohrer, Ebert, & Sibert, 1998). Each potential theme is assigned a fixed direction, so that documents with similar themes will have similar overall shapes.

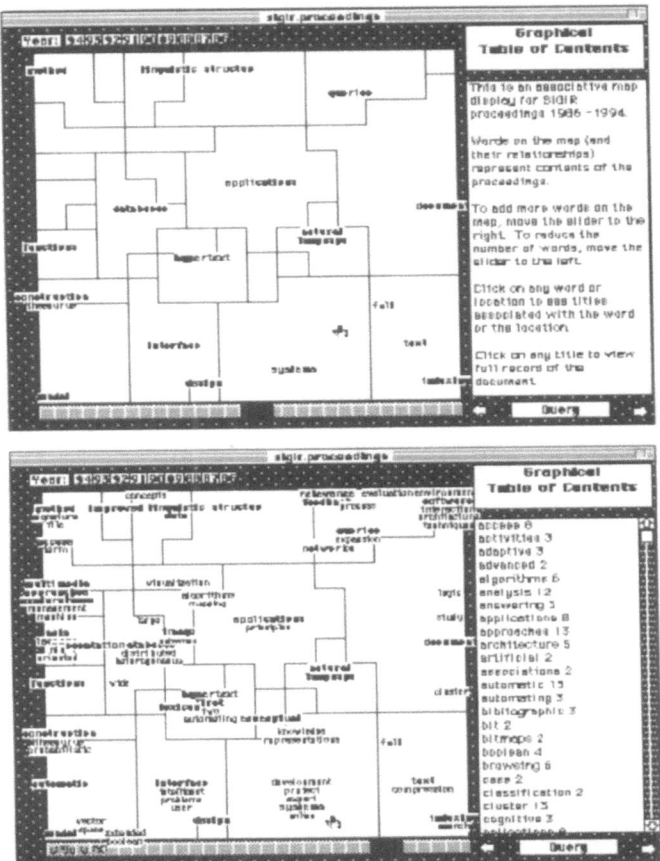

Figure 8. Kohonen map showing term importance and association in a personal collection.
Courtesy of Xia Lin.

Another example is provided in the Envision system, which uses color, shape, and position along the x- and y-axes to convey document similarity (Nowell, France, Hix, Heath, & Fox, 1996). Users are able to assign or change the mapping between graphical attribute and document characteristics to show the particular aspects of similarity in which they are most interested.

6. VISUALIZING LINKS AND CITATIONS

Explicit links can be used for browsing and searching a document collection. Such links can be used for several other purposes as well, including identifying groups of documents that tend to co-cite each other, inferring which are the fundamental or groundbreaking publications, or generally understanding the pattern of links among

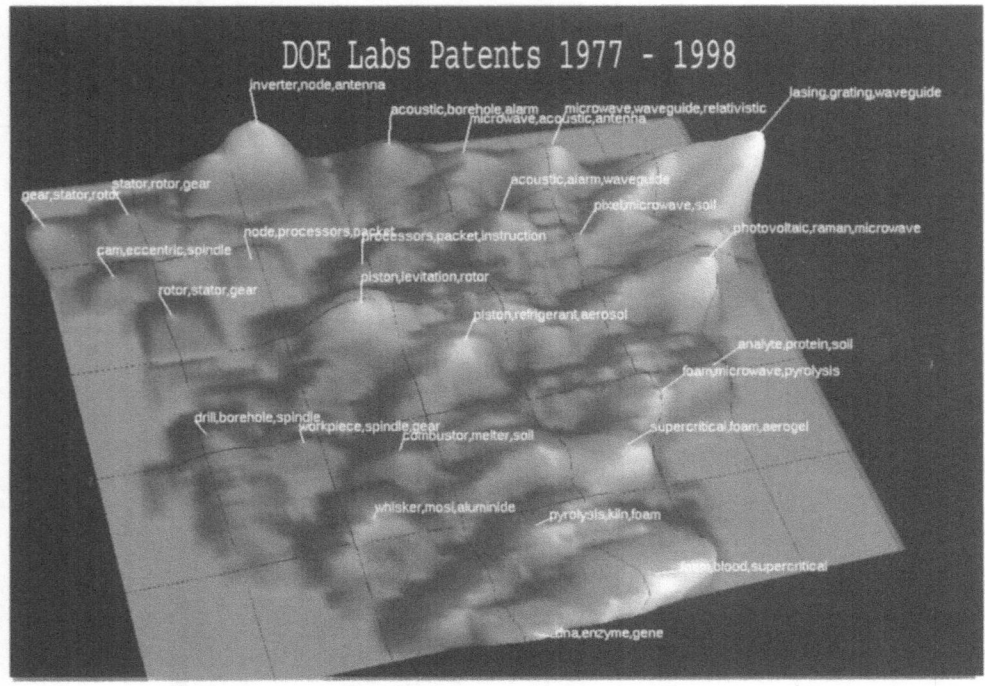

Figure 9. In the ThemeView visualization, the height of the hills indicates thematic strength; proximity indicates relatedness.

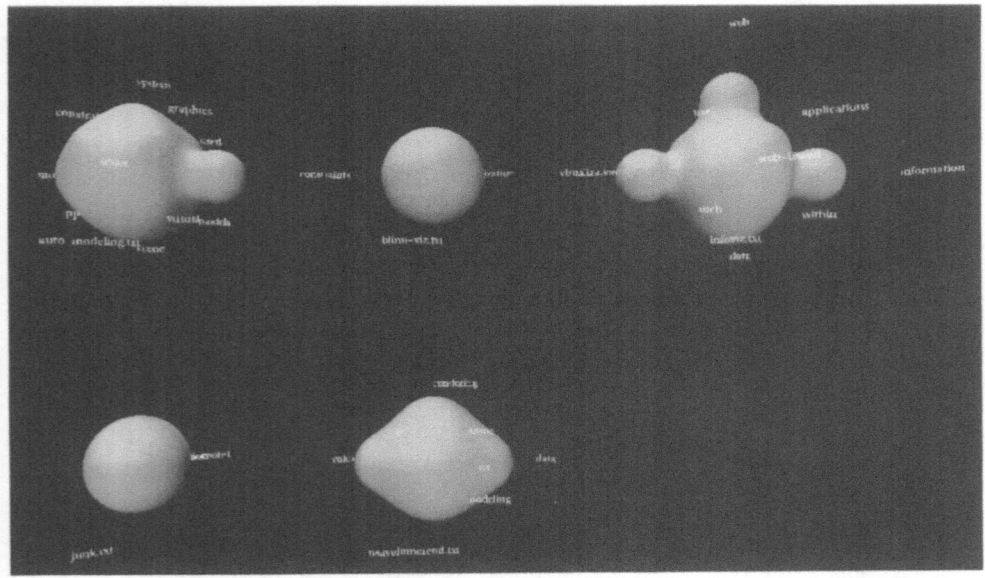

Figure 11. Similar shapes used to convey thematic similarity.
© 1998 IEEE. Reprinted by permission.

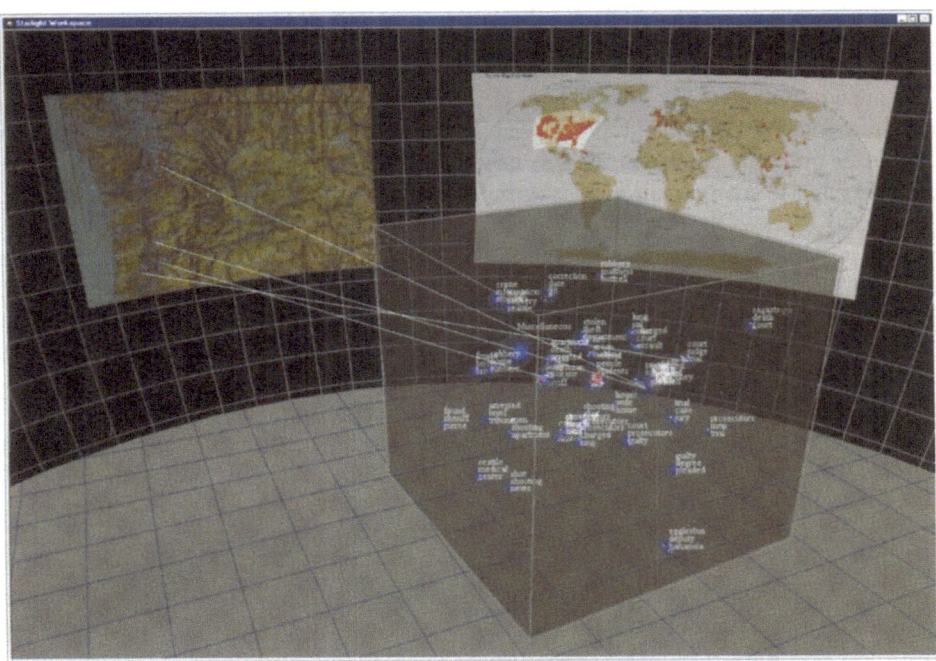

Figure 10. Starlight uses 3D-similarity portrayal and shows linkages, such as common locations mentioned within documents. Image courtesy of Pacific Northwest National Laboratoy.

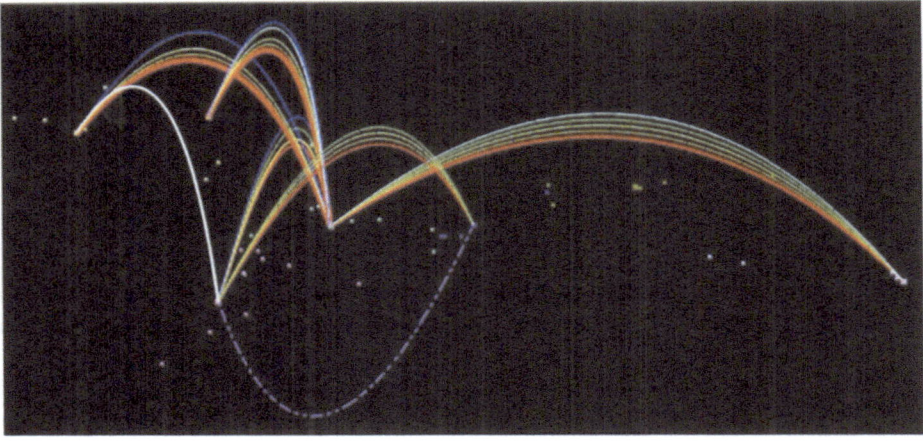

Figure 14. In Rainbows, positive associations are shown as arcs above the plane; strong disassociations are shown as dotted arcs beneath the plane.

documents (Small, 1999). For such exploration and discovery questions, visualization can offer significant advantages.

Explicit link relationships are somewhat different from those previously discussed. They are not transitive, nor do they have degrees of strength; an article either cites another or it does not. However, aggregates can add degrees of strength, for example, where the number of citations among authors is used as a measure of strength. Citations can be viewed as one-to-many relationships (article A cites B and C) or as several binary ones (A cites B and A cites C).

One possible task that visualization could support is to start with a particular document and follow the paths either to articles it references or to articles that cite it. A visualization designed for this task is the Butterfly (MacKinlay, Rao, & Card, 1995), which emphasizes the current document by showing it as the head of a butterfly. The butterfly wings portray the referenced and citing articles respectively (see fig. 12). The system includes a number of features designed to support this kind of citation exploration task.

Another possible task is to understand large patterns of citations or hypertext links. For example, the user may want to identify landmark publications, by understanding which ones are highly cited across various subfields. One temptation is to use a simple directed graph to show all such links, but the result can become highly complex and difficult to use. Fairchild, Poltrock, and Furnas (1988) discuss three approaches to position graph entities to reduce complexity and help illustrate the structure of the information space: basing positions on properties of the entities, using the connectivity itself, and letting users assign position. Chen and Czerwinski (1998) discuss a method for positioning the nodes and also reducing the links actually displayed to emphasize the strongest paths. Figure 13 shows an example portraying author co-citations in the hypertext field (Chen & Carr, 1999). Authors making generic contributions tend to be found near the center of the layout, while those along the outside likely have made contributions in specific areas.

Another approach to visualizing co-citations is provided by Small (1999), who uses co-citations aggregated at various levels to map new and existing scientific subfields based on literature citation patterns.

7. VISUALIZING THE LACK OF RELATIONSHIPS

Previous examples have shown various ways of visualizing the presence of relationships among entities. However, in some cases, the lack of relationships may be equally important. For example, Swanson (1993) has identified examples where disjointed sets of publications contained critical information that, taken together, could lead to a new treatment for a disease or condition. One of the steps in his process relies on verifying the fact that the bodies of literature *do not* cite each other, indicating that the potential combinations have not already been thoroughly explored. Many of the previous visualizations portray such a lack of relationship by the absence of explicit cues, for example, the absence of lines or the fact that articles are remote from each other in the layout. However, it is much easier for people to notice the presence of visual cues than to notice their absence. One system that addresses this issue is Rainbows (see fig. 14). It uses explicit arcs among entities to show relationships. Entities are represented as dots positioned on a plane. Arcs can show presence or absence of relationships. If the

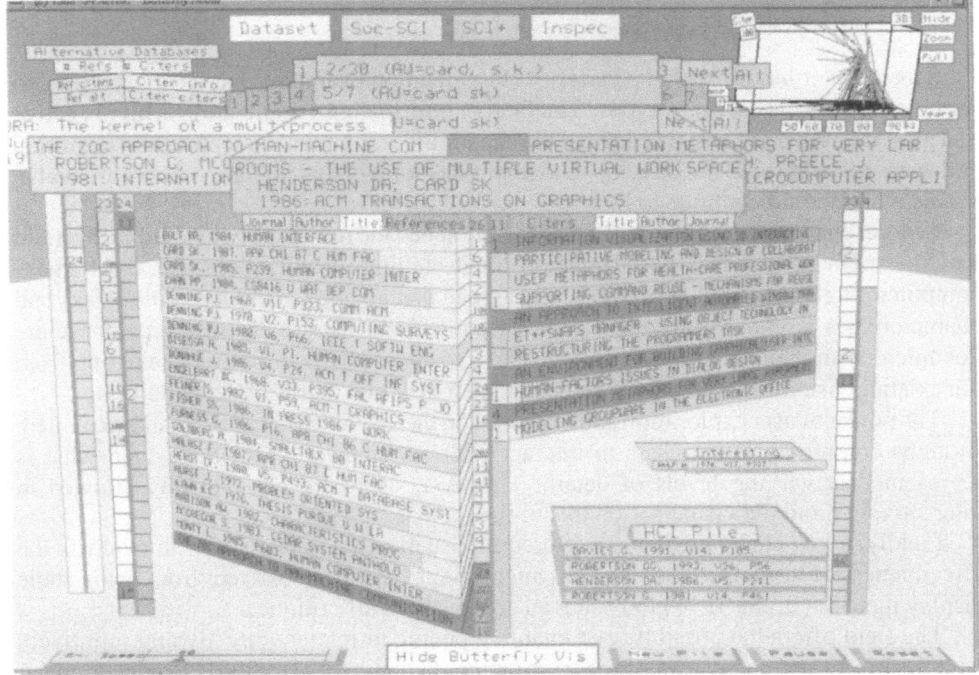

Figure 12. The "wings" of the butterfly show the referenced and citing articles. Mackinlay, Rao, & Card (1995). © 1995 Association for Computing Machinery. Reprinted by permission.

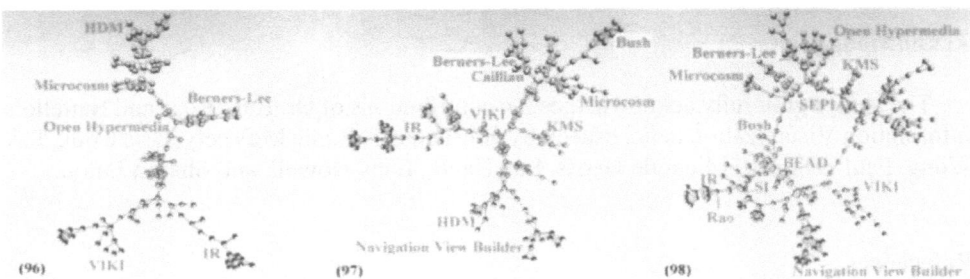

Figure 13. Author co-citations in the hypertext field in various years. ©1999 IEEE. Reprinted by permission.

relationships have continuous strength values, thresholds can be used. Strong associations are shown as arcs above the plane; associations not present or present so weakly that they become disassociations are shown as dotted arcs below the plane. Colors denote the type of association (Hetzler, Harris, Havre, & Whitney, 1998).

8. SUMMARY

This chapter has presented a variety of visualization examples that have been used to portray relationships of many types. The field of information visualization is highly active and growing. As more is learned about the capabilities of human perception and cognition and as more techniques are developed to apply that knowledge, the importance and value of this work should continue to increase.

Several challenges must be addressed to help speed the research. For example, the field would benefit from a task-based method for designing visualizations. In the human-computer interaction field, it is commonly agreed that an understanding of user tasks and characteristics is vital to provide a good user interface. However, even once these are documented and understood, it is not clear how to approach the process of selecting from the existing visualization methods or designing a new one that is better suited.

The power of user exploration and influence on the visualization is also underexploited. Systems commonly allow users to interact with a visualization to understand what is represented at varying levels of detail. However, users are not typically allowed to interactively mold the visualization to suit their needs.

Finally, a key challenge being explored in visualization today is how to evaluate the effectiveness of visualizations. Various suggestions have been offered, from using static display metrics (Brath, 1997) to basing an approach on art critiques.

The field offers the possibility of enabling people to interact directly and intuitively with their information, instead of with a computer (Thomas et al., 2000). However, many visualization systems currently exist only as research prototypes or as systems in very limited use. One of the big challenges in the field is to bring these prototypes more quickly and more effectively to the large groups of users who can potentially benefit from them (Shneiderman, 1999). Work to date, while exciting and creative, is barely scratching the surface of the possibilities.

Acknowledgments

The author gratefully acknowledges the contributions of Dr. Russ Rose and Battelle's Information Visualization team, especially Jim Thomas, Renie McVeety, Kris Cook, Pak Wong, Paul Whitney, Michelle Harris, Sue Havre, Lucy Nowell, and Sharon Eaton.

References

Ahlberg, C., & Shneiderman, B. (1994). Visual information seeking tight coupling of dynamic query filters with starfield displays. *Proceedings on Human Factors in Computing Systems* (SIGCHI '94), 313-317.

Ashcraft, M. (1989). *Human Memory and Cognition*. New York: Harper Collins.

Bertrand-Gastaldy, S. (1986). Improved design of graphic displays in thesauri through technology and ergonomics. *Journal of Documentation*, 42, 225-251.

Brath, R. (1997). Concept demonstration: Metrics for effective information visualization. *Proceedings of IEEE Symposium on Information Visualization* (InfoVis '97), 108-111.

Brodbeck, D., Chalmers, M., Lunzer, A., & Cotture, P. (1997). Domesticating Bead: Adapting an information visualization system to a financial institution. *Proceedings of IEEE Symposium on Information Visualization* (InfoVis '97), 73-80.

Card, S., MacKinlay, J., & Shneiderman, B. (Eds.). (1999). *Readings in Information Visualization*. San Francisco: Morgan Kaufmann.

Chen, C. (1999). *Information Visualization and Virtual Environments*. London: Springer Verlag.

Chen, C., & Carr, L. (1999). Visualizing the evolution of a subject domain: A case study. *Proceedings of IEEE Visualization '99*, 449-452, color plate 561.

Chen, C., & Czerwinski, M. (1998). Latent semantics to spatial hypertext—An integrated approach. *Proceedings of the Ninth ACM Conference on Hypertext and Hypermedia: Links, Objects, Time and Space—Structure in Hypermedia Systems*, 77–86.

Christ, R. (1975). Review and analysis of color coding research for visual displays. *Human Factors*, 17, 542-570.

Cleveland, W., & McGill, R. (1984). Graphical perception: Theory, experimentation, and application to the development of graphical methods. *Journal of the American Statistical Association*, 79, 531-554.

Fairchild, K., Poltrock, S., & Furnas, G. (1988). SemNet: Three-dimensional graphic representation of large knowledge bases. In R. Guindon (Ed.), *Cognitive Science and its Applications for Human-Computer Interaction*. Hillsdale, NJ: Erlbaum.

Havre, S., Hetzler, B., & Nowell, L. (2000). ThemeRiver: Visualizing theme changes over time. *Proceedings of IEEE Symposium on Information Visualization* (InfoVis 2000), 115-123.

Hearst, M., & Karadi, C. (1997). Cat-a-Cone: An interactive interface for specifying searches and viewing retrieval results using a large category hierarchy. *Proceedings of the 20th Annual International ACM SIGIR Conference on Research and Development in Information Retrieval*, 246-255.

Hetzler, B., Harris, M., Havre, S., & Whitney, P. (1998). Visualizing the full spectrum of document relationships. *Structures and Relations in Knowledge Organization, Proceedings of the Fifth International ISKO Conference*, 168-175.

Hoffman, D. (1998). *Visual Intelligence*. New York: W. W. Norton.

Jeong, C., & Pang, A. (1998). Reconfigurable disc trees for visualizing large hierarchical information space. *Proceedings of IEEE Information Visualization* (InfoVis '98), 19-25.

Johnson, B., & Shneiderman, B. (1991). Tree-maps: A space filling approach to the visualization of hierarchical information structures. *Proceedings of IEEE Information Visualization Conference*, 284-291.

Kohonen, T. (1997). Exploration of very large databases by self-organizing maps. *Proceedings of the 1997 IEEE International Conference on Neural Networks* (ICNN '97), PL1-PL3.

Kumar, V., & Furuta, R. (1999). Visualization of relationships. *Proceedings of the Tenth ACM Conference on Hypertext and Hypermedia*, 137-138.

Lakoff, G., & Johnson, M. (1980). *Metaphors We Live By*. Chicago: University of Chicago Press.

Lamping, J., & Rao, R. (1996). The hyperbolic browser: A focus + context technique for visualizing large hierarchies. *Journal of Visual Languages and Computing*, 7, 35-55.

Lancaster, F. (1972). *Vocabulary Control for Information Retrieval*. Washington: Information Resources Press.

Lin, X. (1997). Map displays for information retrieval. *Journal of the American Society for Information Science*, 48, 40-54.

Mackay, W., & Beaudouin-Lafon, M. (1998). DIVA: Exploratory data analysis with multimedia streams. *Proceedings on Human Factors in Computing Systems* (SIGCHI '98), 416-423.

Mackinlay, J., Rao, R., & Card, S. (1995). An organic user interface for searching citation links. *Proceedings on Human Factors in Computing Systems* (SIGCHI '95), 67-73.

Michotte, A. (1963). *The Perception of Causality*. (T. Miles & E. Miles, Trans). New York: Basic Books.

Munzner, T. (1997). H3: Laying out large directed graphs in 3D hyperbolic space. *Proceedings of IEEE Symposium on Information Visualization* (InfoVis '97), 2-10.

Nowell, L. (1997). *Graphical Encoding for Information Visualization: Using Icon Color, Shape, and Size to Encode Nominal and Quantitative Data*. Doctoral dissertation, Virginia Tech. Available: <http://scholar.lib.vt.edu/theses/available/etd-111897-163723/> [2001, October 9].

Nowell, L., France, R., Hix, D., Heath, L., & Fox, E. (1996). Visualizing search results: some alternatives to query-document similarity. *Proceedings of the 19th Annual International ACM SIGIR Conference on Research and Development in Information Retrieval*, 67-75.

Plaisant, C., Heller, D., Li, J., Shneiderman, B., Mushinlin, R., & Karat, J. (1998). Visualizing medical records with LifeLines. *Summary from Human Factors in Computing Systems* (SIGCHI '98), 28-29.

Plaisant, C., Milash, B., Rose, A., Widoff, S., & Shneiderman, B. (1996). Lifelines: visualizing personal histories. *Proceedings on Human Factors in Computing Systems* (SIGCHI '96), 221-227.

Risch, J., Rex, D., Dowson, S., Walters, T., May, R., & Moon, B. (1997). The STARLIGHT information visualization system. *Proceedings 1997 IEEE International Conference on Information Visualization* (IV '97), 42-49.

Rohrer, R., Ebert, D., & Sibert, J. (1998). The shape of Shakespeare: Visualizing text using implicit surfaces. *Proceedings of IEEE Symposium on Information Visualization* (InfoVis '98), 121-129, color plate 160.

Seber, G. (1984). *Multivariate Observations*. New York: John Wiley & Sons.

SGI. (1999). Silicon Graphics MineSet™ Supporting the Discovery Research Process. Available: <http://www.sgi.com/chembio/resources/mineset/index.html> [2001, October 9].

Shneiderman, B. (1999). Crossing the information visualization chasm: From innovation to adoption. Keynote address to *IEEE Symposium on Information Visualization* (InfoVis '99).

Small, H. (1999). Visualizing science by citation mapping. *Journal of the American Society for Information Science*. 50, 799-813.

Sprenger, T. C., Gross, M. H., Bielser, D., Strasser, T. (1998). IVORY: An object-oriented framework for physics-based information visualization in Java. *Proceedings of IEEE Symposium on Information Visualization* (InfoVis '98), 79-86, color plate 155.

Stasko, J., & Zhang, E. (2000). Focus+context display and navigation techniques for enhancing radial, space-filling hierarchy visualizations. *Proceedings of IEEE Symposium on Information Visualization* (InfoVis 2000), 57-65.

Swanson, Don R. (1993). Intervening in the life-cycles of scientific knowledge. *Library Trends,* 41, 606-631.

Technisch Documentatie-en Informatie-Centrum voor de Krijgsmacht [TDCK]. (1963). TDCK Circular Thesaurus System. The Hague: TDCK.

Thomas, J., Cook, K., Crow, V., Hetzler, B., May, R., McQuerry, D., McVeety, R., Miller, N., Nakamura, G., Nowell, L., Whitney, P., & Wong, P. (2000). Human computer interaction with global information spaces—Beyond data mining. In J. Vince & R. Earnshaw (Eds.), *Digital Media: The Future,* 32-46. London: Springer Verlag.

Tufte, E. (1997). *Visual Explanations: Images and Quantities, Evidence and Narrative.* Cheshire, CT: Graphics Press.

Ware, C., Neufeld, E., & Bartram, L. (1999). Visualizing causal relations. *Proceedings of Late Breaking Hot Topics/IEEE Symposium on Information Visualization* (InfoVis '99), 39-42.

Wills, G. & Dill, J. (Eds). (1998). *Proceedings of IEEE Symposium on Information Visualization* (InfoVis '98).

Wills, G. & Keim, D. (Eds). (1999). *Proceedings of IEEE Symposium on Information Visualization* (InfoVis '99).

Wise, J., Thomas, J., Pennock, K., Lantrip, D., Pottier, M., Schur, A., & Crow, V. (1995). Visualizing the non-visual: Spatial analysis & interaction with information from text documents. *Proceedings of IEEE Symposium on Information Visualization* (InfoVis '95), 51-58, color plate 140.

Wong, P., Whitney, P., & Thomas, J. (1999). Visualizing association rules for text mining. *Proceedings of IEEE Symposium on Information Visualization* (InfoVis '99), 120-123, color plate 152.

Index

Information Knowledge and Science Management

Kluwer Academic Publishers – Dordrecht / Boston / London

CO-diffusion ... Nesting and Nursery Management ...

... in the ... Lambert ... W.C. ... "Intermedius Start Log and Track ...
... in ... a ... Snyder ... page 1996 ...   ISBN 978-0-7506-6340
... in ... Main ... name ... sequences of Knowledge ...
... ISBN 1-56-3681-94

... for ... Lambert ... 1998 ... PT., Sequences of Stock Building. An
... in ... for ...   ISBN 1-402-0435-2 7

... in ... Main ... London ... at ... and ... Margaret ...